AF576002

Ship Lifecycle

Ship Lifecycle

Special Issue Editors

Peilin Zhou

Byongug Jeong

MDPI • Basel • Beijing • Wuhan • Barcelona • Belgrade • Manchester • Tokyo • Cluj • Tianjin

Special Issue Editors
Peilin Zhou
University of Strathclyde
UK

Byongug Jeong
University of Strathclyde
UK

Editorial Office
MDPI
St. Alban-Anlage 66
4052 Basel, Switzerland

This is a reprint of articles from the Special Issue published online in the open access journal *Journal of Marine Science and Engineering* (ISSN 2077-1312) (available at: https://www.mdpi.com/journal/jmse/special_issues/Ship_Lifecycle).

For citation purposes, cite each article independently as indicated on the article page online and as indicated below:

LastName, A.A.; LastName, B.B.; LastName, C.C. Article Title. *Journal Name* **Year**, *Article Number*, Page Range.

ISBN 978-3-03936-252-3 (Pbk)
ISBN 978-3-03936-253-0 (PDF)

Contents

About the Special Issue Editors

Peilin Zhou, Professor of Marine Engineering, University of Strathclyde. For over 30 years, Prof. Zhou has dedicated himself to marine technology education and research, which includes maritime LNG systems. He has been actively coordinating and participating in projects funded by EU, UK-EPSRC, and industry. He has published over 180 publications in international journals and conferences on various topics in the areas of marine engineering and marine environmental protection, life cycle analysis, and shipyard process and shipping.

Byongug Jeong, Teaching Associate in Marine Engineering, University of Strathclyde. Dr Jeong is a former sea-going marine engineer and deputy senior marine surveyor for various ships, including LNG carriers. His expertise lies in ship operation, safety/environmental regulations. He has been engaged in several academic and industrial projects. He has also contributed to submitting several agenda and information documents related to the safety and environmental impacts of marine fuels and systems to the International Maritime Organization for future regulatory frameworks.

Editorial

Ship Lifecycle

Peilin Zhou and Byongug Jeong *

Department of Naval Architecture, Ocean & Marine Engineering, University of Strathclyde, 100 Montrose Street, Glasgow G4 0LZ, UK; peilin.zhou@strath.ac.uk

* Correspondence: byongug.jeong@strath.ac.uk

Received: 3 April 2020; Accepted: 4 April 2020; Published: 7 April 2020

Keywords: life cycle assessment (LCA), maritime environment; sustainable production and shipping; CO_2 emissions; NO_x emissions; SO_x emissions; fuel cell

---

With growing concerns of marine pollution, the International Maritime Organization (IMO) has recently adopted a new Resolution MEPC.304 (72), presenting a strategy on curbing greenhouse gas emissions (GHGs) from shipping. Along with this, a series of stringent regulations to limit emissions from shipping activities has been produced at both the international and local level. Such ambitious regulatory works urge us to trust that cleaner production and shipping is one of the most urgent issues in the marine industry.

In order to contribute to global efforts by addressing the marine pollution from various emission types, this Special Issue of the Journal of Marine Science and Engineering was inspired to provide a comprehensive insight for naval architects, marine engineers, designers, shipyards, and ship-owners who strive to find optimal ways to survive in competitive markets by improving cycle time and capacity to reduce design, production, and operation costs while pursuing zero emission.

In this context, this Special Issue is devoted to providing an insight into the latest research and technical developments of ship systems and operation with a life cycle point of view. The goal of this Special Issue is to bring together researchers from across the entire marine and maritime community into a common forum to share cutting-edge research on cleaner shipping. It is strongly believed that such a joint effort will contribute to enhancing the sustainability of marine and maritime activities.

Six novel publications have been dedicated to this Special Issue. First of all, as a proactive response to transitioning to cleaner marine fuel sources, the excellence of the fuel-cell based hybrid ships in several aspects was demonstrated through three publications. Jeon et al. [1] investigated the technical applicability of a molten carbonate fuel cell (MCFC), which is applicable for medium and large-sized ships by means of actual experiment on a hybrid test bed with combined power sources: a 100 kW MCFC, a 30 kW battery bank, and a 50 kW diesel generator. Research outputs demonstrated the technical reliability of MCFC applications on large vessels. Jeon et al. [2] focused on evaluating the safety and reliability of fuel cell-based hybrid power systems applicable for large ships. They adopted the failure mode and effects analysis (FMEA) method with risk priority number (RPN) to evaluate the potential risk of fuel cell systems, providing guidance on the proper approaches into the safety evaluation of marine fuel cells. Roh et al. [3] estimated the economic and environmental impacts of a fuel-cell system. Experiments with the test bed with the hybrid power system were conducted. While applying actual operating conditions for ocean-going ships, fuel consumption, CO_2 emission reduction rates of the hybrid, and conventional power sources were measured. The analysis results from the data of several merchant ships in operation have now revealed the sensitivity of different operating modes on the actual electrical power consumption. The CO_2 emissions of the hybrid system was compared with the case of the diesel generator alone operating in each load scenario where an average of 70%~74% reduction for both fuel consumption and CO_2 emissions was concluded. This research confirmed the excellence of fuel-cells being used as ship power systems. In addition, Jeon et al. [4]

introduced an excellent Active-Front-End (AFE) rectifier applicable for a large electric propulsion ship system. Through a series of simulation, they confirmed the technical maturity of the AFE rectifier as well as the feasibility of the system.

Two publications demonstrated the application of life cycle assessment (LCA) for case studies. Hwang et al. [5] performed a comparative analysis between the conventional diesel fuel oil and liquefied natural gas (LNG) for a 50,000 dead weight tonnage (DWT) bulk carrier, which was the world's first LNG-fueled bulk carrier. Studies have shown that the emissions levels for LNG cases are significantly lower than for MGO cases in all potential impact categories. Gualeni et al. [6] introduced LCA for ship maintenance as a performance assessment (LCPA) tool, which could allow an integration of design with the evaluation of both costs and environmental performances on a comparative basis. An examination of both of these studies concluded that life cycle assessments could provide a better understanding of the overall emissions levels contributed by marine fuel from cradle-to-grave. These conclusions would address the shortcomings of current maritime emission indicators.

From cradle-to-grave, a ship is engaged in various activities, leading to cost investment, energy consumption, and emissions production. This Special Issue has broadly dealt with various aspects of cleaner ship performance and will offer insights into the marine industry with an LCA approach for the application of sustainable energy in marine power systems.

Conflicts of Interest: The authors declare no conflict of interest.

References

1. Jeon, H.; Kim, S.; Yoon, K. Fuel Cell Application for Investigating the Quality of Electricity from Ship Hybrid Power Sources. *J. Mar. Sci. Eng.* **2019**, *7*, 241. [CrossRef]
2. Jeon, H.; Park, K.; Kim, J. Comparison and Verification of Reliability Assessment Techniques for Fuel Cell-Based Hybrid Power System for Ships. *J. Mar. Sci. Eng.* **2020**, *8*, 74. [CrossRef]
3. Roh, G.; Kim, H.; Jeon, H.; Yoon, K. Fuel Consumption and CO2 Emission Reductions of Ships Powered by a Fuel-Cell-Based Hybrid Power Source. *J. Mar. Sci. Eng.* **2019**, *7*, 230. [CrossRef]
4. Jeon, H.; Kim, J.; Yoon, K. Large-Scale Electric Propulsion Systems in Ships Using an Active Front-End Rectifier. *J. Mar. Sci. Eng.* **2019**, *7*, 168. [CrossRef]
5. Hwang, S.; Jeong, B.; Jung, K.; Kim, M.; Zhou, P. Life Cycle Assessment of LNG Fueled Vessel in Domestic Services. *J. Mar. Sci. Eng.* **2019**, *7*, 359. [CrossRef]
6. Gualeni, P.; Flore, G.; Maggioncalda, M.; Marsano, G. Life Cycle Performance Assessment Tool Development and Application with a Focus on Maintenance Aspects. *J. Mar. Sci. Eng.* **2019**, *7*, 280. [CrossRef]

Journal of
Marine Science and Engineering

Article

Large-Scale Electric Propulsion Systems in Ships Using an Active Front-End Rectifier

Hyeonmin Jeon [1], Jongsu Kim [1] and Kyoungkuk Yoon [2,*]

1 Department of Marine System Engineering, Korea Maritime and Ocean University, 727 Taejong-ro, Busan 49112, Korea; jhm861104@kmou.ac.kr (H.J.); jongskim@kmou.ac.kr (J.K.)
2 Department of Electricity, Ulsan Campus of Korea Polytechnics, 155 Sanjeon-gil, Ulsan 44482, Korea
* Correspondence: kkyoon70@kopo.ac.kr; Tel.: +82-10-5541-0424

Received: 1 May 2019; Accepted: 30 May 2019; Published: 1 June 2019

Abstract: In the case of the electric propulsion system on the vessel, Diode Front End (DFE) rectifiers have been applied for large-sized ships and Active Front End (AFE) rectifiers have been utilized for small and medium-sized ships as a part of the system. In this paper, we design a large electric propulsion ship system using AFE rectifier with the proposed phase angle detector and verify the feasibility of the system by simulation. The phase angle derived from the proposed phase angle detection method is applied to the control of the AFE rectifier instead of the zero-crossing method used to detect the phase angle in the control of the conventional AFE rectifier. We compare and analyze the speed control, Direct Current (DC)-link voltage, harmonic content and measurement data of heat loss by inverter switch obtained from the simulation of the electric propulsion system with the 24-pulse DFE rectifier, the conventional AFE rectifier, and the proposed AFE rectifier. As a result of the simulation, it was confirmed that the proposed AFE rectifier derives a satisfactory result similar to that of a 24-pulse DFE rectifier with a phase shifting transformer installed according to the speed load of the ship, and it can be designed and applied as a rectifier of a large-sized vessel.

Keywords: electric propulsion system; DFE rectifier; AFE rectifier; phase angle detector

1. Introduction

As environmental pollution has become a global issue, the International Maritime Organization (IMO) has been strengthening regulations on emissions of sulfur oxides, nitrogen oxides, and carbon dioxide from ships [1,2]. As a result of that various researches are being carried out in order to cope with environmental regulations that are strengthening internationally in the shipbuilding and shipping industries [3]. Moreover, the electric propulsion system of vessels with propulsion motors is also one of emerging countermeasures [4–7]. As shown in Figure 1, the order of environmentally friendly electric propulsion ship is dramatically increased on 2017 World Fleet Resister by Clarkson's Research [8].

The components of the conventional large-sized electric propulsion ship are generally composed of generator, DFE rectifier with phase shifting transformer, inverter and propulsion motor, and it is possible to design the size of the engine room with some margin [9–11]. In an electric propulsion ship, when the switching of inverters occurs, a harmonic current is generated in a power system [12]. Thus, large and small problems occur in the generator, transformer, and propulsion motor. Various methods for reducing harmonics have been studied. In the case of large electric propulsion systems, phase shifting transformer has been adopted as the most common method of installing a transformer on the output side of a generator [3,13]. There are various methods of harmonics reduction of the DFE rectifier using a phase shifting transformer, such as multi-pulse of the rectifier output [14–16], active filter installation [17], and improvement of the transformer connection method [18–20].

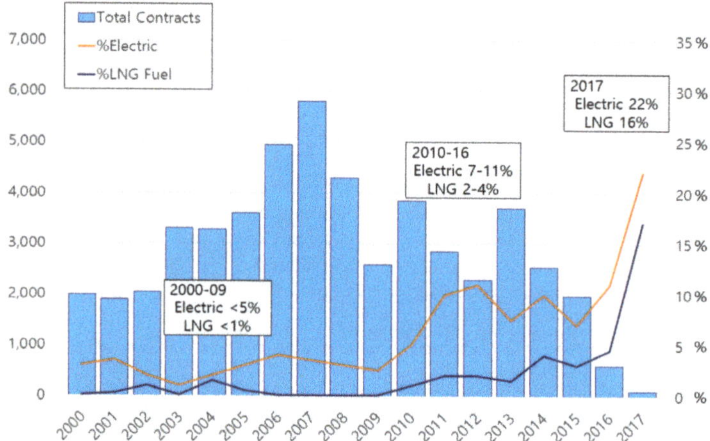

Figure 1. Annual ship newbuilding contracts.

However, when the phase shifting transformer is installed, there is a disadvantage as installation space and cost increase. Moreover, it is difficult to apply it to a small and medium-sized ship with limited space. AFE rectifiers have been mainly applied to small and medium-sized electric propulsion ships [21,22], but recently, as the technologies of power semiconductors with high capacity and high speed switching characteristics have been developed, so that it is possible to model a large-sized electric propulsion system using AFE rectifier [23]. The AFE rectifier must be designed with a control circuit that can control the semiconductor switch, and it is especially necessary to accurately detect the phase angle of the power supply voltage. Zero crossing technique that can detect the phase angle quickly is simple and has no special control method [24–27]. However, due to the fluctuation of generator output voltage in case of high load, such as propulsion motor or bow thruster. The detection of the phase angle may not be performed momentarily. Various methods have been studied to overcome the severe disadvantage of this zero-crossing technique.

In large-sized commercial vessels, it is crucial to secure the space for cargo transportation as much as possible [3,9]. However, to reduce the harmonics contained in the ship power system, the DFE rectifiers with large-sized phase shifting transformer have a disadvantage to load cargo. And the AFE rectifier using the existing zero-crossing technique also has various problems [28,29].

In this paper, an AFE rectifier using the Phase Locked Loop (PLL) method is applied to a large electric propulsion system instead of the phase angle detection method using the zero-crossing method [30–33]. We used the power analysis program, Power Simulation (PSIM), to model an AFE rectifier that uses the PLL method. Comparison simulations were performed for large-scale electric propulsion systems with the conventional DFE as well as proposed AFE rectifiers. Based on this simulation, the resulting speed of the propulsion motor, DC output of the DC link, and harmonic output characteristics of the input power supply were analyzed based on the type of rectification. In addition, the thermal loss in the switching element, which is present in the inverter when AFE rectifiers are used, as well as its stability, were evaluated. Based on these results, the characteristics of the DFE and AFE rectifiers in a large-scale electric propulsion system were compared to confirm the higher effectiveness of using the PLL-method-based AFE rectifiers in large-scale electric propulsion systems compared with the use of conventional DFE rectifiers.

2. Conventional Methods for Marine Electrical Prolusion Systems

2.1. Background

The DFE rectifier with a phase shifting transformer is mostly used for high-power drives, such as the motors, fans, and compressors installed in the large plant in the industrial field [34]. Thanks to its long history of operation with know-how and track records accumulated in the industrial field, it was proved that stable operation would be possible. Therefore, the same high-power drive system of the existing industrial field has been applied in the early large electric propulsion system [6].

As mentioned in the previous section, thus far, large-scale electric propulsion systems have primarily consisted of a generator, phase shifting transformer, DFE rectifier, inverter, and propulsion motor [35]. The generators that supply power to the large-scale electric propulsion systems are typically brushless synchronous generators that can generate high voltages, such as 3300 V or 6600 V. To reduce the detrimental effects of the aforementioned harmonics produced in these power systems, including on the voltage and current of the generator, and to improve the output waveform of the rectified DC current, a phase shifting transformer is installed before the DFE rectifier [36,37]. Furthermore, to control the speed of the propulsion motor, an inverter which can control voltage and frequency is installed. Induction motors are often used as propulsion motors because it is easy to control the torque and speed of such motors. In addition, their maintenance is simple [38,39]. Figure 2 shows the schematic diagram of a large-scale electric propulsion system.

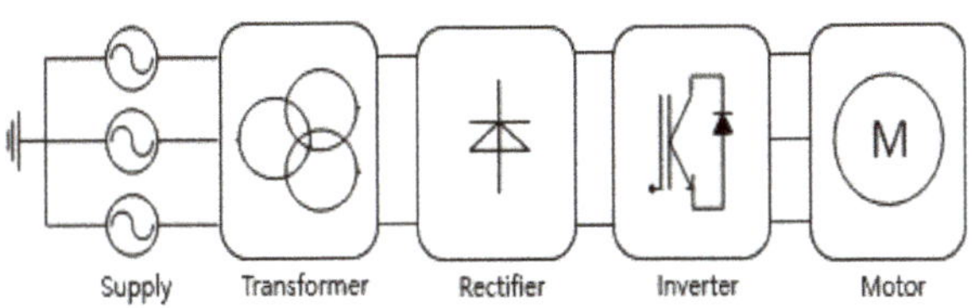

Figure 2. Schematic diagram of a large-scale electric propulsion system.

Although contributing to decreasing the total harmonics distortion, the DFE rectifier with the transformer is subject to the increase of volume and weight for the system as well as the design complexity for the phase shifting transformer to obtain a linear DC waveform by increasing the number of pulses. For example, Figure 3 shows the electric drawing of the electric propulsion system of 'A' company with a 24-pulse rectifier. Table 1 shows the comparison when an AFE rectifier is installed instead of a 24-pulse rectifier [40]. The total volume and weight of the system will be increased inevitably. Figure 4 shows the actual application of the transformer installed on the ship.

Table 1. Comparison of AFE rectifier vs. 24-pulse rectifier.

Component	24-Pulse DFE Rectifier	AFE Rectifier
Propulsion Motor	2 pcs 41,679 kg × 2 = 83,358kg	2 pcs 41,679 kg × 2 = 83,358 kg
Phase Shifting Transformer	4 pcs 11,940 kg × 4 = 47,760kg	-
Rectifier	8pcs 4730 kg × 8 = 37,840kg	4pcs 3760 kg × 4 = 15,040 kg
Inverter	4 pcs 3760 kg × 4 = 15,040kg	4 pcs 3,760 kg × 4 = 15,040 kg
Total Weight	183,998 kg	113,438 kg

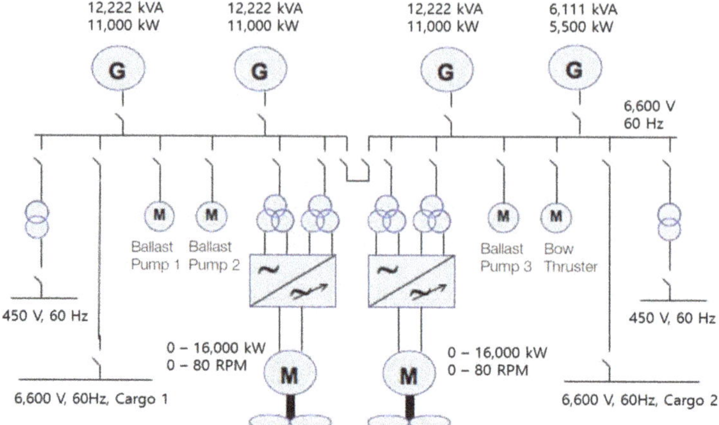

Figure 3. Typical configuration for a twin skeg electric propulsion ship.

Figure 4. Phase shifting transformer installed on large-scale electric propulsion system.

2.2. DFE Rectification Method in Large-Scale Electric Propulsion Systems

Electric propulsion systems require the AC current generated by the generator to be converted into DC current. The conventional method involves the use of a DFE rectifier that employs a diode element to generate 6-pulses. However, as shown in Table 2 below, 6-pulse rectifiers lead to high harmonic distortion so that it cannot suit ship application. In order to reduce the harmonic distortion, as shown in Figure 3, a typical electric propulsion system maker chooses the phase shifting transformer to reduce the level of harmonics distortion by making 12-pulse, 24-pulse DC output [14,15]. Therefore, taking into account the harmonics distortion in the existing electric propulsion system, complex structures of phase shifting transformer have been applied with a number of DFE rectifiers. As a result, not only does the initial installation cost increase but also the volume and weight of the system [35].

Consequently, the overall efficiency of the system is reduced because of this decrease in the input power factor, as well as the severely distorted waveforms, owing to the DC output in a pulse form [41]. To resolve these problems, a high-capacity passive filter and phase shifting transformer must be installed, which, in turn, have their own drawbacks in that they considerably increase the overall system size and installation costs associated with the system [9].

Table 2. Total harmonic distortion for type of rectifier

Rectifier Type	Total Harmonic Distortion (THD)
6-Pulse	25~27%
12-Pulse	8~11%
18-Pulse	4~5%
24-Pulse	2~3%
AFE	4~5%

2.2.1. 12-Pulse Rectifier

One rectification method for improving the harmonic characteristics of the output of the power supply is to install a phase shifting transformer before the DFE rectifier to produce DC waveforms with 12-pulses in each cycle [15]. Figure 5 shows the block diagram of a DFE-style 12-pulse rectifier that uses a phase shifting transformer. The connections on the transformer's secondary side consist of Y-Y and Y-D connections. The phase shift angle for creating 12-pulses per cycle is 30° between each phase, as indicated by Equation (1).

$$\Delta = \angle e_{ab} - \angle e_{AB} = 30^\circ \quad (1)$$

where, Δ is the phase shift angle, $\angle e_{ab}$ is the line voltage of the primary side of the rectifier, and $\angle e_{AB}$ is the line voltage of the secondary side of the rectifier.

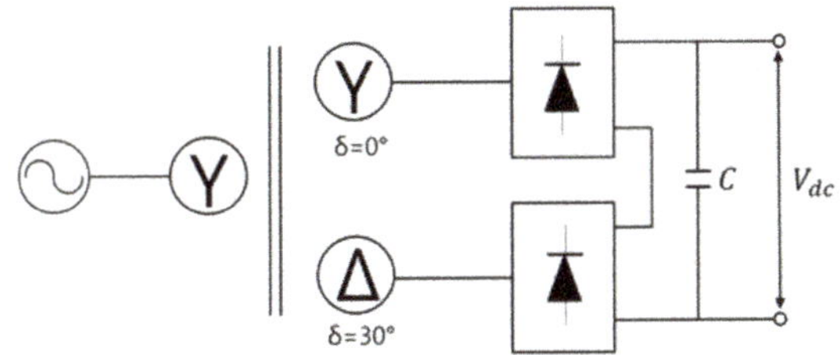

(**a**) Block diagram of 12-pulse DFE rectifier.

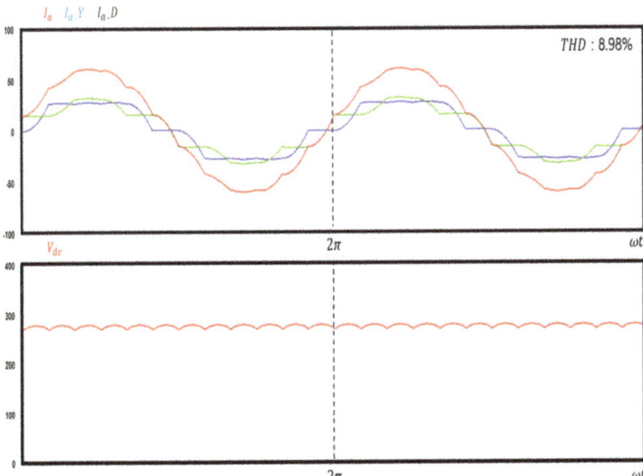

(**b**) Current and DC link Voltage of 12-pulse rectifier.

Figure 5. Block diagram and Waveforms of a DFE-style 12-pulse rectifier.

Thus, in this case, for the 12-pulse DC waveforms that occur during one cycle, there is a 30° difference in the phases of the Y and D connections on the secondary side of the phase shifting transformer. In addition, there is a 30° difference in the phases of the 6-pulse DC waveforms generated by each unit, which produce a 12-pulse-per-cycle DC waveform in the DC link unit. In terms of the total harmonic distortion in the DC output waveform of the 12-pulse rectifier, the fifth order and seventh order harmonics are entirely eliminated, and only the harmonics that are 11th order and above remain. Thus, the rectifier effectively reduces the harmonic characteristics more effectively compared with a 6-pulse rectifier.

2.2.2. 24-Pulse Rectifier

A 24-pulse rectifier uses a zig-zag-shaped phase shifting transformer to create 24-pulse waveforms per cycle. Compared with the 12-pulse rectification method described above, this rectifier can produce better DC voltage waveforms and reduce harmonics more effectively. Figure 6 shows a block diagram of the 24-pulse rectifier. The phase shift angle of the phase shifting transformer, in this case, can be expressed via Equation (2) specified below, in particular, there is a phase difference of 15°.

$$\Delta = \angle e_{ab} - \angle e_{AB} = 15^\circ \tag{2}$$

where, Δ is the phase shift angle, $\angle e_{ab}$ is the line voltage of the primary side, and $\angle e_{AB}$ is the line voltage of the secondary side.

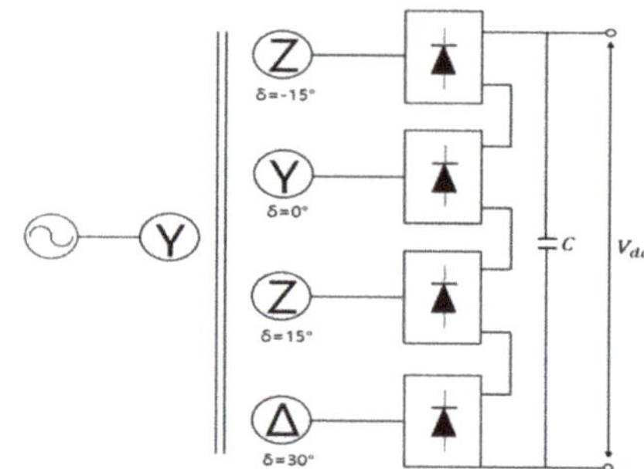

(**a**) Block diagram of 24-pulse DFE rectifier.

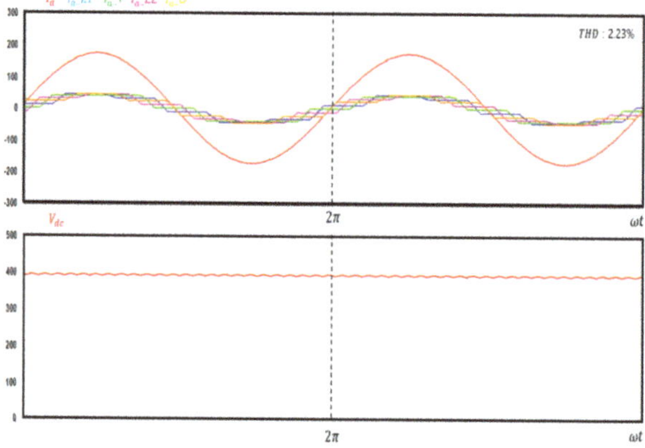

(**b**) Current and DC link Voltage of 24-pulse rectifier.

Figure 6. Block diagram and Waveforms of a 24-pulse rectifier.

Considering the total harmonic distortion in the output waveforms generated by the 24-pulse rectifier, all lower-order harmonics below the 19th order are eliminated. Therefore, the 24-pulse rectifier has significantly better harmonic output characteristics than those of the 12-pulse rectifier. In addition, because the waveforms of the DC link unit include 24-pulses per cycle, this leads to a voltage waveform that is considerably similar to DC voltage.

2.3. AFE Rectification Method in Large-Scale Electric Propulsion Systems

The AFE rectifier uses semiconductor-based technologies, such as Insulated Gate Bipolar Transistor (IGBTs), Integrated Gate Commutated Thyristor (IGCTs), and Metal Oxide Semiconductor Field Effect Transistor (MOSFETs), among others, which can turn power semiconductor elements off and on as required. Based on the control style of the semiconductor element, power conversion may be realized automatically. In particular, a fixed DC output voltage can be maintained even if the load changes. Thus far, AFE rectifiers have been primarily used in small- to mid-sized electric propulsion systems on ships owing to the limited capacities of the power semiconductor elements in these rectifiers [42].

The AFE rectifier must continuously measure the supply voltage to control the rectifier. As shown in Figure 7, the error of phase angle, which is crucial for the control of the rectifier, occurs momentarily due to the deterioration of the voltage quality, such as harmonics and noise included in the supply voltage.

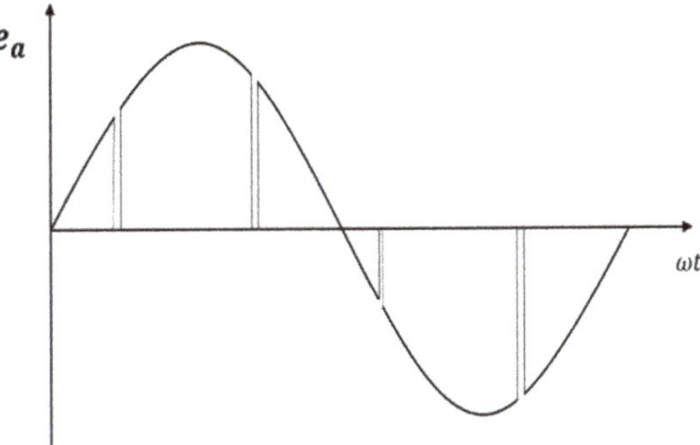

Figure 7. An example of phase angle error.

As shown in Figure 8, the form of the AFE rectifier is the same as that of an inverter that converts DC current to AC current. The AFE rectifier consists of a total of three units and six power semiconductor switches. In addition, it includes an inductor that controls the input current of the power supply, as well as a capacitor that maintains a fixed DC output voltage.

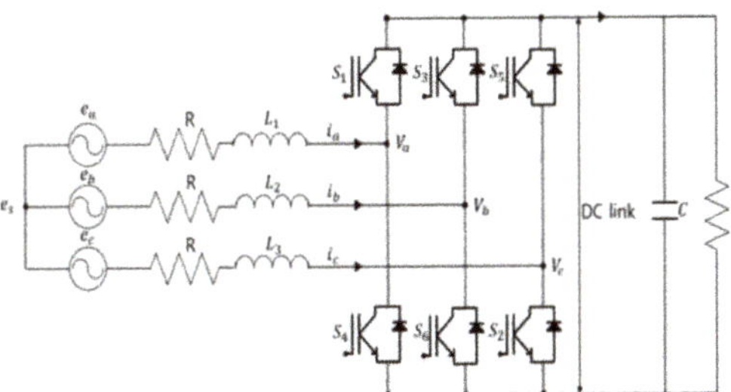

Figure 8. Block diagram of an AFE rectifier.

Equation (3) is the voltage equation of the AFE rectifier.

$$e_{abc} = Ri_{abc} + L\frac{di_{abc}}{dt} + V_{abc} \quad (3)$$

where, e_{abc} is the three-phase power supply voltage, i_{abc} is the phase current, and V_{abc} is the input side voltage of the rectifier.

The AFE rectifier controls the level and phase of the AC input current, i_s, while performing the power conversion. The AFE rectifier must control the level of the voltage that is applied to the inductor on the input side. In particular, i_s is controlled by controlling the input voltage of the rectifier, i.e., V_{rec}. Figure 9 shows the equivalent circuit for an AFE rectifier. The voltage V_L that is applied to the inductor can be obtained using Equations (4) and (5).

$$e_s = V_{Rec} + V_L \quad (4)$$

$$V_L = \omega L i_s \quad (5)$$

where, e_s is the AC input to the power supply, V_L is the inductor voltage, and V_{rec} is the rectifier input voltage.

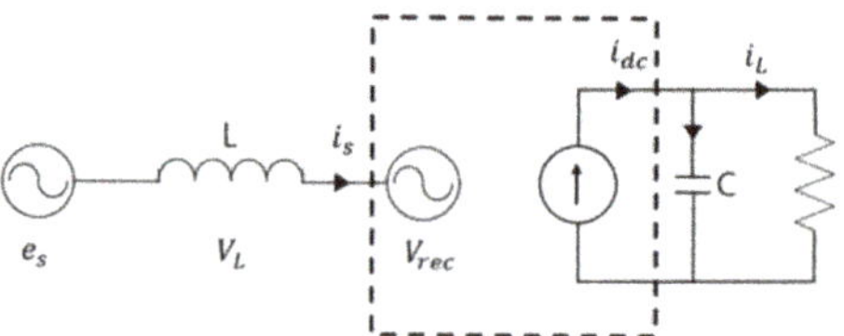

Figure 9. Equivalent circuit of the AFE equivalent circuit.

In order to control the AFE rectifier, it is necessary to find the d-q axis coordinate and current reference values that are in phase with the power supply voltage. To find these values, it is necessary to find the phase angle θ. In the conventional AFE rectifiers, the zero-crossing technique is used to find the phase angle θ for control. The zero-crossing technique involves measuring the power supply voltage and finding the 0 values that occur at each half cycle to estimate the current phase angle θ. Figure 10 shows the block diagram of a phase angle detector that uses the zero-crossing technique.

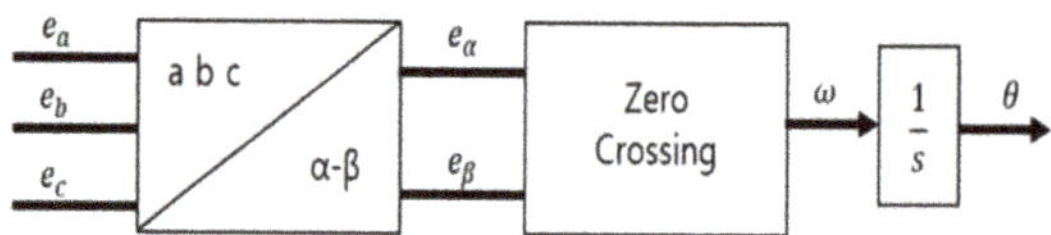

Figure 10. Phase angle detector using the zero-crossing technique.

In particular, to find the phase angle, the moment when the power supply voltage changes from negative to positive can be set as the standard angle 0°, as depicted in Figure 11 Alternatively, the three-phase AC power supply values can be converted to a static coordinate system to find the phase angle directly, as given by Equation (6).

$$\theta = \tan^{-1}\left(\frac{e_\alpha}{e_\beta}\right) \quad (6)$$

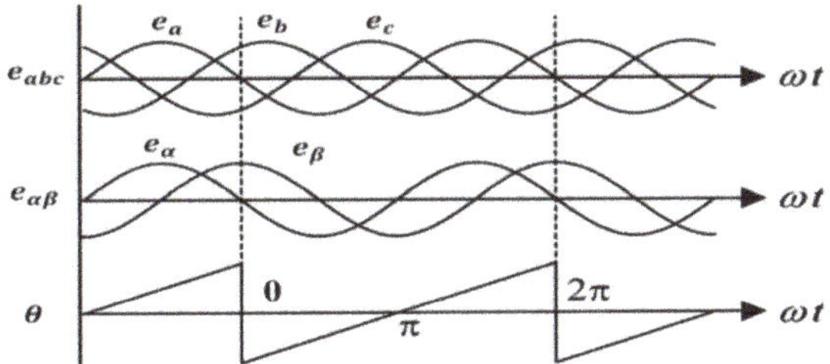

Figure 11. Relationship between the power supply voltage and phase angle in the case of the zero-crossing technique.

One advantage of the zero-crossing technique is that it can be used to find the phase angle in a simple manner. However, in some cases, some zero points are missed during the phase detection step, and consequently, its estimation speed is slow. Furthermore, another disadvantage of the zero-crossing technique is that estimation errors occur when noise due to harmonics or voltage notching occurs. Therefore, in this study, we created an AFE rectifier control circuit that uses the PLL method to accurately find the phase angle.

3. Large-Scale Electric Propulsion System Using an Improved AFE Rectifier

With the recent development of high-capacity power semiconductor elements, which can be used in large-scale electric propulsion systems, it has become theoretically possible for AFE rectifiers, which, thus far, were primarily used in small- to mid-sized electric propulsion systems on ships, to be used in large-scale electric propulsion systems.

3.1. Improved AFE Rectifier Control

As can be deduced from the voltage equation of the AFE rectifier, i.e., Equation (3), the three-phase AC voltage and current values continuously change as time progresses. Therefore, it is difficult to ensure stable control over the rectifier. To control the rectifier in a simple, yet accurate manner, it is necessary to convert the coordinates of the three-phase AC power supply to a given standard axis in order to convert the voltage equation of the AFE rectifier to that with a stable DC value. In particular, by changing the coordinate system using such a conversion, the three-phase AC levels, which continuously change over time, can be converted to two DC values d-q, which are easy to control. These converted values can then be used to control the AFE rectifier [22,33].

Figure 12 shows the structure of a phase angle detector that uses the PLL method rather than the existing zero crossing technique to find the phase angle in an AFE rectifier. The PLL phase angle detector converts the voltage of the three-phase AC power supply to a value on the d-q axis in a synchronous rotating coordinate system. These values can then be used to find voltages e_d, e_q, which are DC voltages, and therefore, easy to control. In our study, e_d is arbitrarily set to be the active power, while e_q is set to be the reactive power. Then, the value of the reactive power e_q can be controlled to ensure that it is 0.

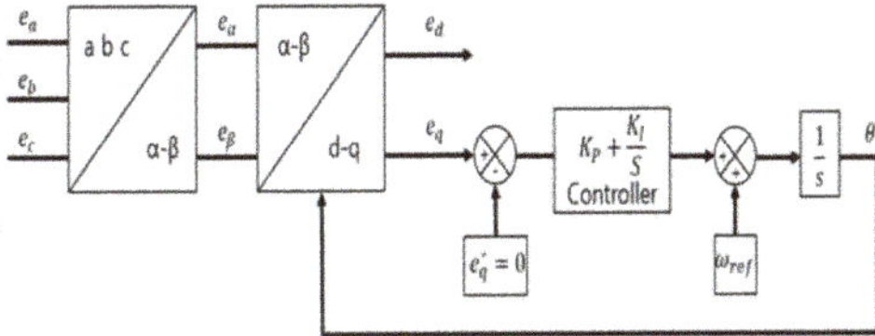

Figure 12. Block diagram of a phase detector using the PLL method.

Figure 13 shows the relationship between the q-axis voltage of the synchronous rotating coordinate system and voltage phase angle for controlling e_q so that it is zero. In Equation (7), P_s is the active power supplied by the power supply.

$$P_s = e_a i_a + e_b i_b + e_c i_c = \frac{3}{2}\left(e_d i_d + e_q i_q\right) = \frac{3}{2} e_d i_d \tag{7}$$

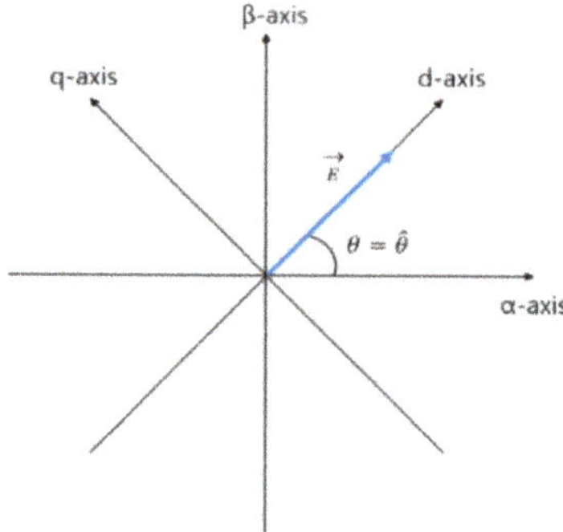

Figure 13. Imaginary phase angle, which is the same as the actual phase angle.

Thus, the power supply current is only affected by e_d, which indicates the voltage value on the d-axis, whereas the current value on the q-axis current does not have any effect on it. Therefore, the supplied current can be given by $P_s = \frac{3}{2} e_d i_d$. If the active power values are set on the d-axis, and the q-axis voltage is controlled so that it is 0, the actual phase angle θ becomes the same as the imaginary phase angle $\hat{\theta}$ as shown in Figure 13. Consequently, the phase angle θ can be accurately determined.

To find the actual phase angle θ, Equations (8) and (9) can first be solved by performing a coordinate conversion on the three-phase e_a, e_b, e_c voltages of the AC power supply so that e_a, e_b, e_c transform into voltages on the $\alpha - \beta$ axis of the static coordinate system, as shown in Figure 14.

$$e_\alpha = E\cos\theta \tag{8}$$

$$e_\beta = E\sin\theta \tag{9}$$

where, E is the peak value of the input AC phase voltage, and θ is the phase angle between a and the α-axis.

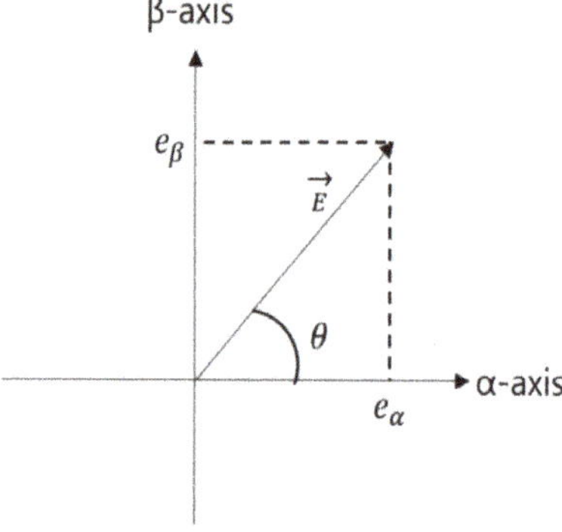

Figure 14. Conversion to input voltage's coordinate axis.

The values that have been converted to the static coordinate system with the $\alpha-\beta$ axis are then converted to a synchronous rotating coordinate system with a $d-q$ axis using Equation (10). This is shown in Figure 15.

$$\begin{bmatrix} e_d \\ e_q \end{bmatrix} = \begin{bmatrix} \cos\theta & \sin\theta \\ -\sin\theta & \cos\theta \end{bmatrix} \begin{bmatrix} e_\alpha \\ e_\beta \end{bmatrix} \tag{10}$$

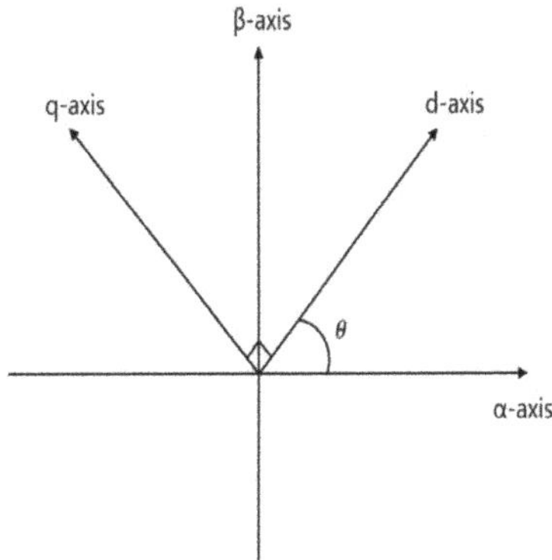

Figure 15. Coordinate conversion to a synchronous rotating coordinate system.

If the phase angle is the imaginary phase angle $\hat{\theta}$ rather than the actual phase angle θ, the voltage values on the $d-q$ axis of the synchronous rotating coordinate system can be expressed using Equation (11) as follows.

$$\begin{bmatrix} e_d \\ e_q \end{bmatrix} = \begin{bmatrix} \cos\hat{\theta} & \sin\hat{\theta} \\ -\sin\hat{\theta} & \cos\hat{\theta} \end{bmatrix} \begin{bmatrix} e_\alpha \\ e_\beta \end{bmatrix}. \tag{11}$$

Equation (11) can be represented as Equations (12) and (13) on simplification. Furthermore, Equations (8) and (9) can be used to obtain the voltage values in the static coordinate system using the actual phase value, and if they are substituted in Equations (12) and (13), the resulting equations are Equations (14) and (15), respectively.

$$e_d = e_\alpha \cos\hat{\theta} + e_\beta \sin\hat{\theta} \tag{12}$$

$$e_q = -e_\alpha \sin\hat{\theta} + e_\beta \cos\hat{\theta} \tag{13}$$

$$e_d = E\left(\cos\theta\cos\hat{\theta} + \sin\theta\sin\hat{\theta}\right) = Ecos\left(\theta - \hat{\theta}\right) \tag{14}$$

$$e_q = E\left(-\cos\theta\sin\hat{\theta} + \sin\theta\cos\hat{\theta}\right) = Esin\left(\theta - \hat{\theta}\right) \tag{15}$$

where, θ is the actual phase value between the α and d axes, whereas $\hat{\theta}$ is the imaginary phase angle.

In Equation (15), if q-axis voltage value e_q is adjusted to be 0, then $\theta = \hat{\theta}$, and Equations (16) and (17) can be solved. Consequently, the imaginary phase angle $\hat{\theta}$ can be controlled so that it matches the actual phase angle θ.

$$e_d = Ecos\left(\theta - \hat{\theta}\right) = E \tag{16}$$

$$e_q = Esin\left(\theta - \hat{\theta}\right) = 0. \tag{17}$$

If the three-phase input current is converted to the $d-q$ coordinate axis system based on the phase angle θ that is accurately obtained with the phase angle detector using the PLL control for the AFE rectifier, the DC equivalent values of the AC current can be obtained. Thus, it is possible to use a proportional integral controller to obtain an AFE rectifier with excellent control performance.

Equations (18) and (19) represent the voltage equation in the synchronous rotating coordinate system with the $d-q$ axes.

$$e_d = Ri_d + L\frac{di_d}{dt} - \omega L i_q + V_d \quad (18)$$

$$e_q = Ri_q + L\frac{di_q}{dt} - \omega L i_d + V_q. \quad (19)$$

As is clear from Equations (18) and (19), there are speed electromotive and counter-electromotive force components that represent a disturbance in the current control. In particular, these components act as mutual interference components in the current control so that changes in the d-axis current affect the q-axis current and vice-versa. Therefore, in order to obtain good control characteristics when current control is performed in a synchronous rotating coordinate system, it is necessary to design a current controller that includes feed-forward compensation for the counter-electromotive and speed electromotive forces as indicated in Figure 16. If a feed-forward controller is included, the output voltage of a proportional integral current controller in a synchronous rotating coordinate system can be expressed as Equations (20) and (21).

$$e_d^* = e_{d-fb}^* + e_{d-ff}^* \quad (20)$$

$$e_q^* = e_{q-fb}^* + e_{q-ff}^* \quad (21)$$

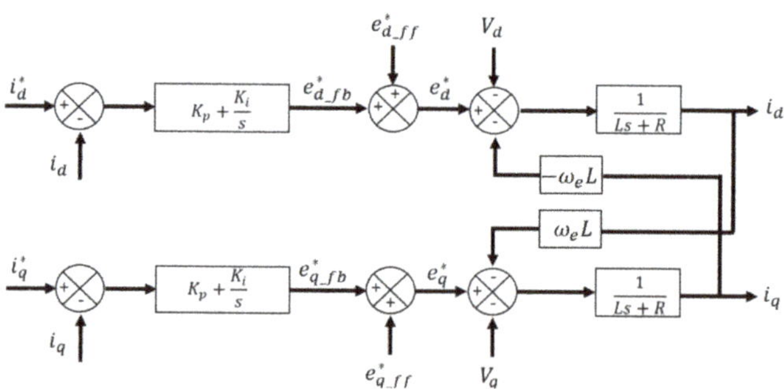

Figure 16. Current controller in the AFE rectifier that compensates for the counter-electromotive force.

3.2. Controlling the Propulsion Motor Speed of a Large-Scale Electric Propulsion System Using the Improved AFE Rectifier

In this study, we used an indirect vector control technique to control the propulsion motor speed of a large-scale electric propulsion system. The indirect vector control technique uses flux current, torque current, and electric motor constant in the synchronous rotating coordinate system to calculate the slip reference angular speed. The integral value of this added to the rotor speed is considered as the flux angle. For high-performance torque and flux control, the stator current that is provided to the motor is divided into different components that match each of the components that are orthogonal to the standard flux [38,43].

In Figure 17, the $\alpha-\beta$ axis is fixed to the stator, while the d–q axis rotates at the synchronous angular speed ω_θ. The rotating axis is matched to the d axis, while the slip angle (θ_{sl}) to the rotor axis is maintained as the axis rotates. Therefore, the stator current supplied to the electric motor is divided into the flux component current i_{ds} and torque component current i_{qs}, which can be used to perform high-quality torque and flux control.

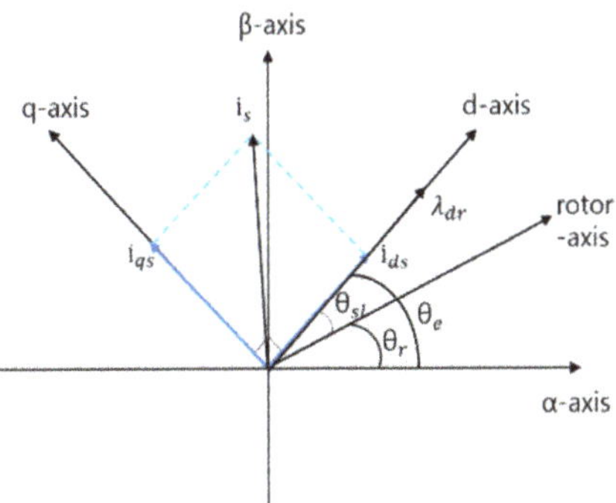

Figure 17. Indirect vector control.

In the case of rotator flux-based indirect vector control, the rotator flux is controlled to ensure that it only exists as a d-axis component. Therefore, Equation (22) is valid.

$$\lambda_{qr} = \rho\lambda_{qr} = 0 \tag{22}$$

As can be deduced from Equation (23), the torque is proportional to i_{qs}. Therefore, i_{qs} can be considered as the torque component current. Furthermore, when flux control is constant, the rotator flux can be controlled via i_{ds}. Therefore, i_{ds} can be considered as the flux component.

$$T_e = \frac{3}{2}\frac{P}{2}\frac{L_m}{L_r}\lambda_{dr}i_{qs} \tag{23}$$

Thus, when i_{ds} is constant, the slip relational equation can be expressed as Equation (24).

$$\omega_{sl} = \frac{R_r}{L_r}\frac{i_{qs}}{i_{ds}} \tag{24}$$

The position of the rotator flux can be obtained as the integral value of the sum of the electric motor speed and slip reference angular speed, as shown in Equation (25) below.

$$\theta_e = \int(\omega_r + \omega_{sl})dt \tag{25}$$

4. Methodology

In the present study, several different large-scale electric propulsion systems were modeled using the DFE rectifier, phase shifting transformer, conventional AFE rectifier, and improved AFE rectifier. A comparative analysis was performed on the operating characteristic results that were obtained for the large-scale electric propulsion system with the improved AFE rectifier and conventional DFE and AFE rectifiers in Figure 18. The modeled large-scale electric propulsion system is the same as the electric propulsion systems that are installed and used in current LNG tankers. The propulsion motor parameters are listed in Table 3. Figure 19 is a graph showing the oscillation of the output voltage of the generator due to the load variation during the actual operation of the ship. Thus, in order to verify the effectiveness of the proposed AFE rectifier, an arbitrary waveform was inserted into the output of the generator during the simulation.

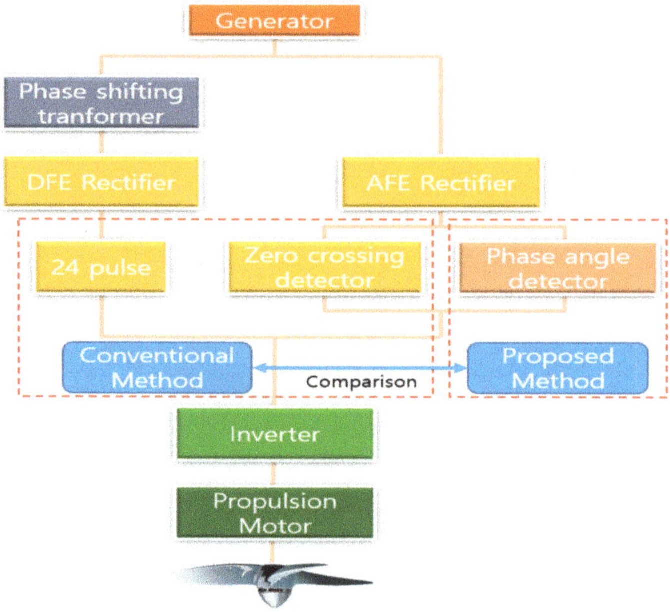

Figure 18. Flowchart of comparison analysis with each rectifier.

Table 3. Propulsion of large-scale electric propulsion ships applied to simulations electric motor parameters.

Item	Value	Item	Value
Rated Power	6000 Kw	R_S	0.0167 Ω
Rated Voltage	3300 V	L_S	1.49 mH
Rated Current	1200 A	R_r	0.07 Ω
Rated Speed	650.8 rpm	L_r	0.35 mH
Frequency	60 Hz	L_m	48 mH
Number of Poles	10	J	169 Kg/m^2

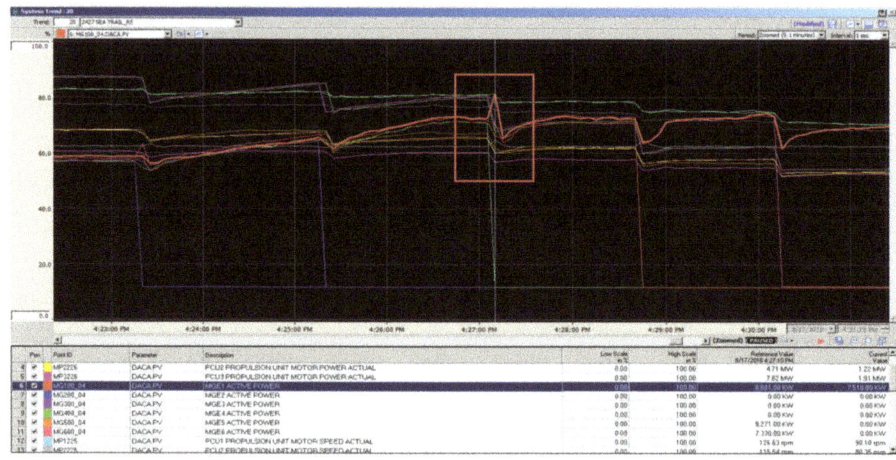

Figure 19. The hunting of the output voltage of generator due to load variation in the actual operation of the ship.

To compare the characteristics of the large-scale electric propulsion systems based on their rectification method, systems were developed that used the improved AFE rectifier in Figure 20 as well as one that used an existing 24-pulse rectifier in Figure 21 with a phase shifting transformer. The improved AFE rectifier and power semiconductor in the inverter used the same IGBT module. The primary specifications of the power semiconductor are listed in Table 4 [44].

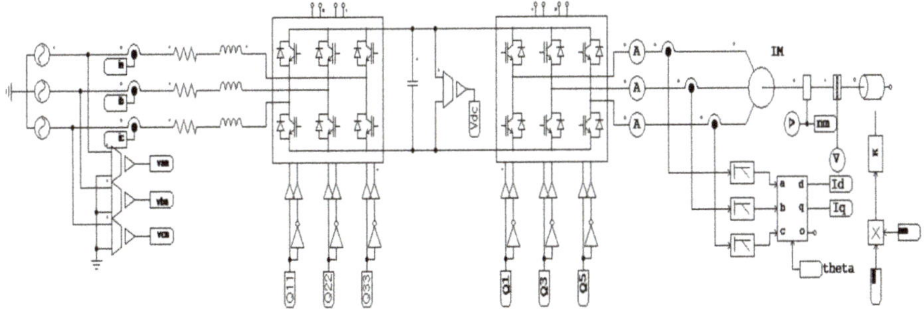

Figure 20. Overall block diagram of the electric propulsion system with the improved AFE rectifier.

Table 4. Semiconductor model and specification for IGBT power used in AFE rectifier.

Model	ABB Hi-Pak 5SNA 1200G450350
Collector-emitter voltage	4500 V
DC collector current	1200 A
Peak collector current	2400 A
DC forward current	1200 A

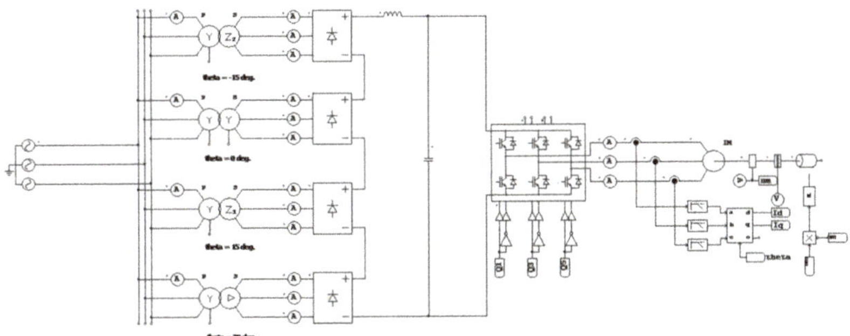

Figure 21. Overall block diagram of the electric propulsion system with a 24-pulse rectifier that employs a phase shifting transformer.

In order to verify the robustness of the proposed AFE rectifier control, an irregular waveform was arbitrarily generated by inserting instantaneous zero-point noise, harmonics, etc. in the output power voltage of the generator during the simulation.

In Figure 20, the phase angle that is needed to control the proposed improved AFE rectifier was obtained using a phase angle detector with a PLL controller, as shown in Figure 22. The q-axis voltage, which is the reactive power component that is converted to a synchronous rotating coordinate system, was controlled always to be 0, so that the phase angle would match the actual phase angle during phase angle determination.

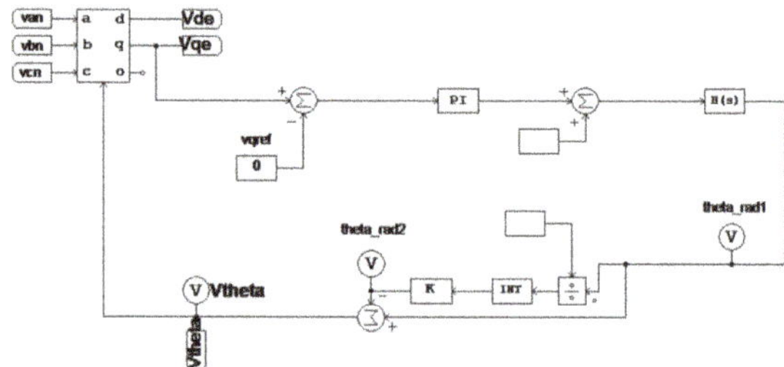

Figure 22. Phase angle detector with a PLL controller.

The current controller for the AFE rectifier is shown in Figure 23. In particular, the output of the DC link was compared with the reference voltage value in real time using the current controller used for controlling the AFE rectifier. Furthermore, the error was controlled by the proportional integral controller, and the feed-forward compensation technique was used to remove the mutual interference components of the counter electromotive and speed electromotive forces.

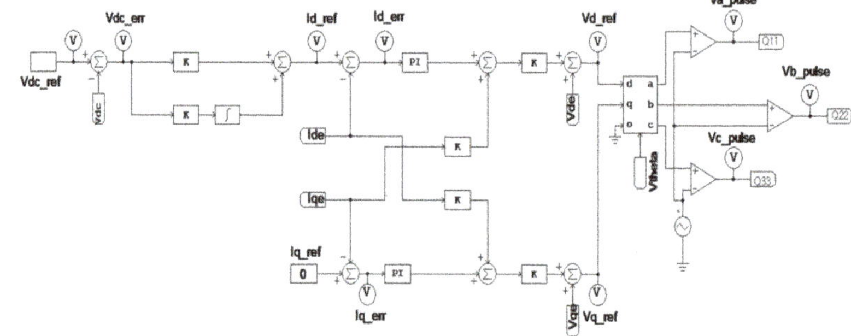

Figure 23. Current controller for the AFE rectifier.

In addition, the inverter used the indirect vector technique to control the speed and torque of the propulsion motor. This indirect vector control is depicted in Figure 24. Simulations of the modeled circuits were performed to compare speed response characteristics of the propulsion motor, DC output voltage waveforms of the DC link, input side harmonic output characteristics of the power supply, and heat loss in the inverter switching element, among others based on the changes in the rectification method of the simulated large-scale electric propulsion system.

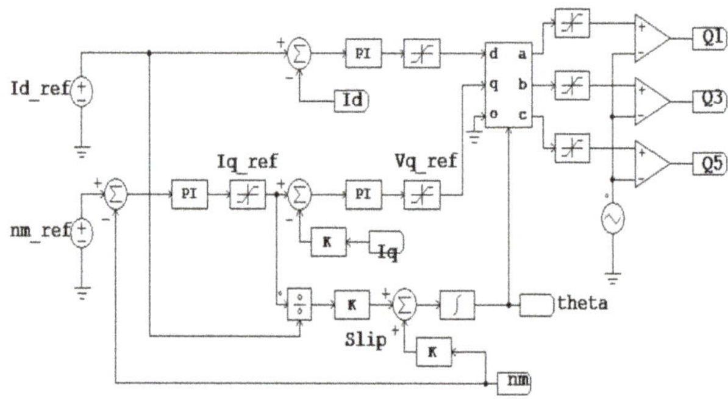

Figure 24. Block diagram of the indirect vector control for propulsion motor speed.

5. Analysis Simulations of the Operating Characteristics of Large-Scale Electric Propulsion Systems According to Rectification Method

5.1. Speed Response Characteristics of the Propulsion Motor

Figure 25 shows the results of the simulation on the characteristics of the propulsion motor's response to step speed reference values. The standard speed value was set as 500 rpm. Based on these results, it can be observed that the conventional AFE rectifier, 24-pulse rectifier, and improved AFE rectifier all showed relatively good speed response characteristics to the reference value.

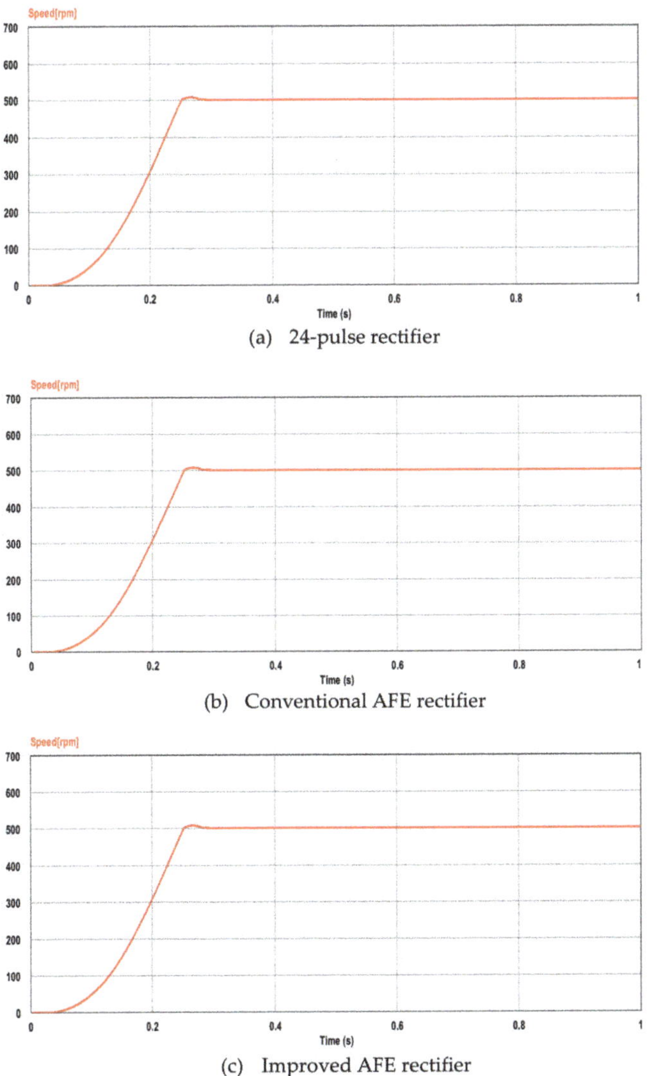

(a) 24-pulse rectifier

(b) Conventional AFE rectifier

(c) Improved AFE rectifier

Figure 25. Speed response characteristics of the propulsion motor.

5.2. Comparison of the DC Output Voltage Waveform in the DC Link

Figure 26 shows the DC output voltage in the DC link when different rectifiers are used. The 24-pulse rectifier, which uses the phase shifting transformer, had the best response characteristics to the reference DC voltage. Nevertheless, the improved AFE rectifier maintained the DC output voltage in a more stable state than the conventional AFE rectifier.

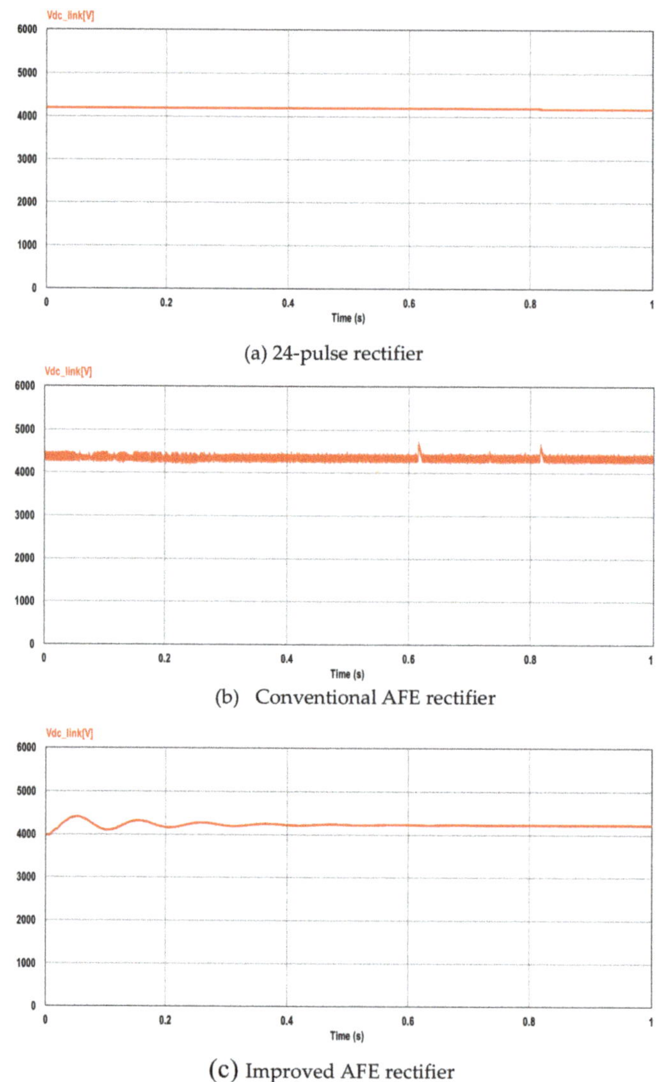

Figure 26. DC output voltage in the DC link.

5.3. Comparison of the Total Harmonic Distortion in the Voltage of the Power Source

Figure 27 shows the results of the simulation for the total harmonic distortion in the power supply at the output side. The estimated values for the electric propulsion system that used the improved AFE rectifier were about 2% better than the case wherein the conventional AFE rectifier was used.

Similar results were obtained for the 24-pulse rectifier. These values satisfied the recommendation for total harmonic distortion that is included in the IEEE Standard 519-2014 [45], which specifies a total harmonic distortion of 8% in power generators under 1 kV.

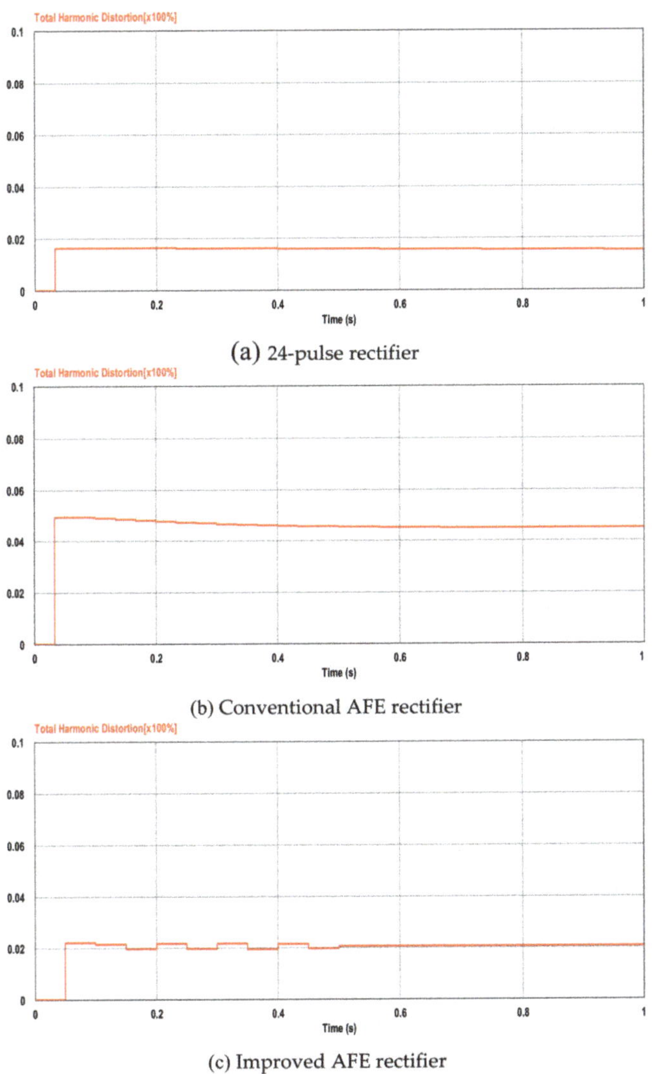

(a) 24-pulse rectifier

(b) Conventional AFE rectifier

(c) Improved AFE rectifier

Figure 27. Total harmonic distortion of the power source on the voltage side.

5.4. Comparison of the Heat Loss in the Inverter Switching Element

The inverter's IGBT module consists of the IGBT element and diode. Loss during power conversion can be categorized into switching loss and conduction loss in the IGBT element as well as switching loss and conduction loss in the diode element. The simulation estimated the power loss and heat loss in the inverter module's diode and IGBT element. The equation for obtaining conduction loss and switching loss is as follows. Figure 28 shows the block diagram of the apparatus used to record heat loss.

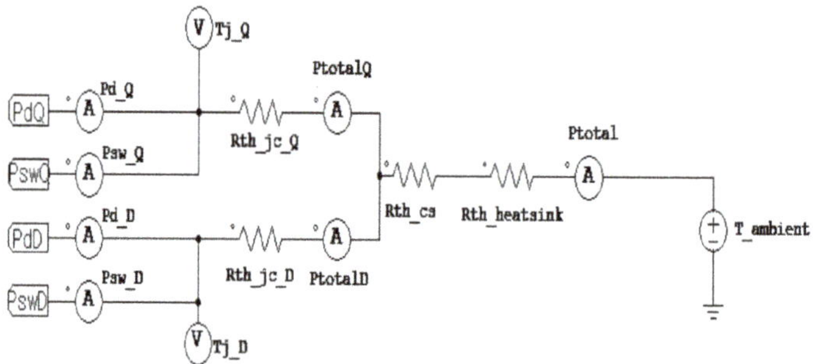

Figure 28. Block diagram of the apparatus used to obtain the heat loss in the inverter switching element.

Conduction losses:

$$P_{cond\_Q} = V_{ce(sat)} \times I_c \times D \tag{26}$$

Switching losses:

$$\text{Turn on } P_{sw\_Q\_on} = E_{on} \times f \times V_{cc} \div V_{cc\_datasheet} \tag{27}$$

$$\text{Turn off } P_{sw\_Q\_off} = E_{off} \times f \times V_{cc} \div V_{cc\_datasheet} \tag{28}$$

where, $V_{ce(sat)}$ is the transistor collector–emitter saturation voltage, I_c is the collector current, D is conducting duty cycle, E_{on} is the transistor turn-on energy losses, E_{off} is the transistor turn-off energy losses, f is the frequency, and V_{cc} is the actual dc bus voltage.

Figure 29 shows the results of our simulations to estimate the sum of the switching and conduction losses of the diode in the inverter of a large-scale electric propulsion system using the conventional AFE rectifier, 24-pulse rectifier, and improved AFE rectifier. Table 5 lists the average of these results. It is clear from these analysis results that the system with the improved AFE rectifier shows lower switching losses than the one with the 24-pulse rectifier.

Table 5. Comparison of the heat losses from the inverter when using the conventional AFE rectifier, DFE rectifier and improved AFE rectifier in a large-scale electric propulsion system modeled.

	Conventional AFE	24 Pulse	Improved AFE
Diode loss average [W]	514	453	445
IGBT loss average [W]	3717	3321	3350
Total loss average [W]	4231	3774	3795

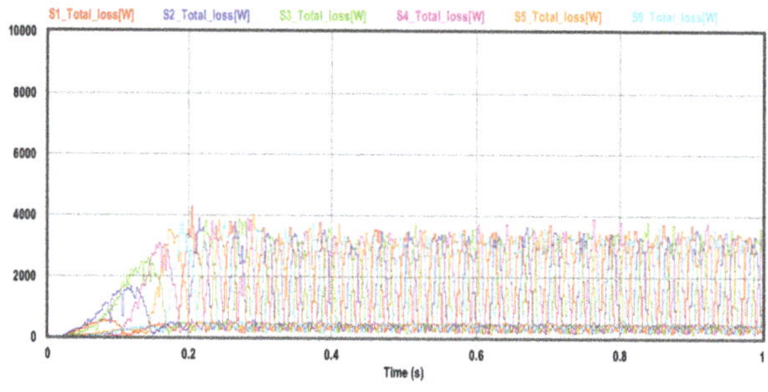

(a) 24-pulse rectifier

Figure 29. *Cont.*

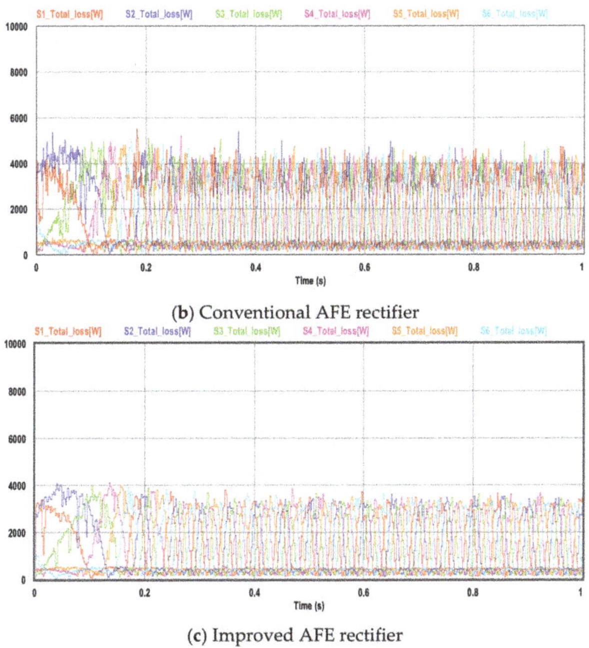

(**b**) Conventional AFE rectifier

(**c**) Improved AFE rectifier

Figure 29. Comparison of the switching loss in the inverter elements.

6. Conclusions

Relatively good propulsion motor speed response results were obtained for the large-scale electric propulsion systems that were modeled using each of the rectification methods. However, the large-scale electric propulsion system that used the improved AFE rectifier was able to accurately detect the phase angle of the power supply voltage using a PLL control circuit. Therefore, it was able to obtain a more stable DC link voltage output and reduced number of harmonics in the input power supply side compared with the conventional AFE rectifier. Furthermore, the improved rectifier showed similar output performance as the 24-pulse DFE rectifier, which uses a phase shifting transformer. In addition, from the simulation results, it was observed that when the proposed AFE rectifier is used, the DC output performance of the DC link in the rectifier was improved. Consequently, the switching loss of the power semiconductor of the inverter used for controlling the speed of the propulsion motor was similar to that of the 24-pulse rectifier.

So far, large-sized electric propulsion vessels have adopted the same high-power drive system of the shore, but the use of phase shifting transformers with DFE rectifiers in a limited space and heavy weight has several disadvantages. In this study, an improved AFE rectifier with superior performance compared to the 24-pulse rectifier was applied. Research findings revealed that the proposed system would mitigate the complexity of the electric converting system by reducing the number of the rectifiers to be fitted. Therefore, the optimized spatial arrangement in the engine room could contribute to increasing the cargo loading efficiency of the vessel.

Based on these simulation results, it was confirmed that it can be more effective for a large-scale electric propulsion system to use an AFE rectifier that can turn a power semiconductor switch on and off, rather than using an existing DFE rectifier employing a phase switching transformer.

Author Contributions: Conceptualization, J.K and K.Y.; Methodology, J.S., K.Y. and H.J.; Formal analysis, H.J.; Software, K.Y. and H.J.; Writing-original draft preparation, H.J.; Writing-review and editing, K.Y.

Funding: This research received no external funding.

Acknowledgments: We thank our colleagues from Seoung Hwan Kim who provided insight and expertise that greatly assisted the research, although they may not agree with all of the conclusions of this paper. In addition, they are indebted towards Prof. Kim for his assistance on setting-up the simulations in the PSIM environment.

Conflicts of Interest: The authors declare no conflict of interest.

References

1. International Maritime Organization. Marine Environment Protection Committee 72nd Session. Available online: http://www.imo.org/en/MediaCentre/MeetingSummaries/MEPC/Pages/MEPC-72nd-session.aspx (accessed on 13 April 2018).
2. International Maritime Organization. Marine Environment Protection Committee. Available online: http://www.imo.org/en/MediaCentre/HotTopics/Pages/Sulphur-2020.aspx (accessed on 23 June 2016).
3. ABB Group. System Project Guide for Passenger. Available online: https://library.e.abb.com/public/608df0ae42ea2ce8c1257abd004ff506/ABB_System_Project_Guide_Passenger_Vessels.pdf (accessed on 1 February 2011).
4. Hansen, J.F.; Wendt, F. History and State of the Art in Commercial Electric Ship Propulsion, Integrated Power Systems, and Future Trends. *Proc. IEEE* **2015**, *103*, 2229–2242. [CrossRef]
5. McCoy, T.J. Trends in Ship Electric Propulsion. In Proceedings of the IEEE Power Engineering Society Summer Meeting, Chicago, IL, USA, 21–25 July 2002; pp. 343–346.
6. Skjong, E.; Cunningham, J.; Johansen, T.A.; Rodskar, E.; Molinas, M.; Volden, R. Past, Present, and Future Challenges of the Marine Vessel's Electrical Power System. *IEEE Trans. Transp. Electrif.* **2016**, *2*, 522–537. [CrossRef]
7. Vasquez, C.A. Methodology to Select the Electric Propulsion System for Platform Supply Vessels. Master's Thesis, Universidade de Sao Paulo, Sao Paulo, Brazil, 2014.
8. Clarksons Research. Available online: https://clarksonsresearch.wordpress.com/tag/propulsion/ (accessed on 31 March 2017).
9. Ajioka, Y. Electric Propulsion Systems for Ships. *Hitachi Rev.* **2013**, *62*, 231–232.
10. Ådnanes, A.K. *Maritime Electrical Installations and Diesel Electric Propulsion*; ABB AS: Oslo, Norway, 2003; 86.
11. McCoy, T.J.; Amy, J.V. The State-of-the-Art of Integrated Electric Power and Propulsion Systems and Technologies on Ships. In Proceedings of the IEEE Electric Ship Technologies Symposium ESTS, Baltimore, MA, USA, 20–22 April 2009; pp. 340–344.
12. Hoevenaars, T.; Evans, I.C.; Lawson, A. New Marine Harmonic Standards. *IEEE Ind. Appl. Mag.* **2010**, *16*, 16–25. [CrossRef]
13. Verboomen, J.; Van Hertem, D.; Schavemaker, P.H.; Kling, W.L.; Belmans, R. Phase shifting transformers: Principles and applications. In Proceedings of the 2005 International Conference on Future Power Systems, Amsterdam, The Netherlands, 18 November 2005.
14. Kim, J.S.; Choi, J.H.; Yoon, K.K.; Seo, D.H. Harmonic Reduction of Electric Propulsion Ship using New Rectification Scheme. *J. Korean Inst. Inf. Commun. Eng.* **2012**, *16*, 2230–2236. [CrossRef]
15. Kim, J.S. Harmonic Reduction of Electric Propulsion Ship by Multipulse Drive. *J. Korean Inst. Marit. Inf. Commun. Sci.* **2010**, *15*, 425–431.
16. Sun, J.; Bing, Z.; Karimi, K.J. Input Impedance Modeling of Multipulse Rectifiers by Harmonic Linearization. *IEEE Trans. Power Electron.* **2009**, *24*, 2812–2820.
17. Akagi, H.; Isozaki, K. A Hybrid Active Filter for a Three-Phase 12-Pulse Diode Rectifier Used as the Front End of a Medium-Voltage Motor Drive. *IEEE Trans. Power Electron.* **2012**, *27*, 69–77. [CrossRef]
18. Kamath, G.R.; Runyan, B.; Wood, R. A Compact Autotransformer Based 12-Pulse Rectifier Circuit. In Proceedings of the 27th Annual Conference of the IEEE Industrial Electronics Society, Denver, CO, USA, 29 November–2 December 2001; pp. 1344–1349.
19. Skibinski, G.L.; Guskov, N.; Zhou, D. Cost Effective Multi-Pulse Transformer Solutions for Harmonic Mitigation in AC Drives. In Proceedings of the 38th IAS Annual Meeting on Conference Record of the Industry Applications Conference, Salt Lake City, UT, USA, 12–16 October 2003; pp. 1488–1497.

20. Wen, J.; Qin, H.; Wang, S.; Zhou, B. Basic Connections and Strategies of Isolated Phase-Shifting Transformers for Multipulse Rectifiers: A Review. In Proceedings of the 2012 Asia-Pacific Symposium on Electromagnetic Compatibility APEMC 2012, Singapore, 21–24 May 2012; pp. 105–108.
21. Kim, S.Y.; Choe, S.; Ko, S.; Kim, S.; Sul, S.K. Electric Propulsion Naval Ships with Energy Storage Modules through AFE Converters. *J. Power Electron.* **2014**, *14*, 402–412. [CrossRef]
22. Kim, S.Y.; Cho, B.G.; Sul, S.K. Consideration of Active-Front-End Rectifier for Electric Propulsion Navy Ship. In Proceedings of the 2013 IEEE Energy Conversion Congress & Expo, Denver, CO, USA, 15–19 September 2013; pp. 13–19.
23. Applied Materials. Available online: http://www.appliedmaterials.com/nanochip/nanochip-fab-solutions/december-2013/power-struggle (accessed on 8 March 2013).
24. Gupta, A.; Rajev, T.; Sachin, M. Review on Zero Crossing Detector. *India IJECT J.* **2012**, *7109*, 366–368.
25. Irmak, E.; Colak, I.; Kaplan, O.; Guler, N. Design and Application of a Novel Zero-Crossing Detector Circuit. In Proceedings of the International Conference on Power Engineering Energy and Electrical Drives, Malaga, Spain, 11–13 May 2011; pp. 1–4.
26. Vainio, O.; Ovaska, S.J. Noise Reduction in Zero Crossing Detection by Predictive Digital Filtering. *IEEE Trans. Ind. Electron.* **1995**, *42*, 58–62. [CrossRef]
27. Wall, R.W. Simple Methods for Detecting Zero Crossing. In Proceedings of the 29th Annual Conference of the IEEE Industrial Electronics Society (IEEE Cat. No.03CH37468), Roanoke, VA, USA, 2–6 November 2003; Volume 3, pp. 2477–2481.
28. Parvez, M.; Mekhilef, S.; Tan, N.M.L.; Akagi, H. An Improved Active-Front-End Rectifier Using Model Predictive Control. In Proceedings of the 2015 IEEE Applied Power Electronics Conference and Exposition (APEC), Charlotte, NC, USA, 15–19 March 2015; pp. 122–127.
29. Chomat, M. Operation of Active Front-End Rectifier in Electric Drive under Unbalanced Voltage Supply. *Electr. Mach. Drives* **2012**, *10*, 195–216.
30. Chung, S.K. Phase-Locked Loop for Grid-Connected Three-Phase Power Conversion Systems. *IEE Proc. Electr. Power Appl.* **2000**, *147*, 213–219. [CrossRef]
31. Trivedi, M.S.; Mujumdar, U.B. Study and Simulation of Parks Transform PLL for robust single phase grid connected system. *Int. J. Recent Innov. Trends Comput. Commun.* **2015**, *3*, 149–152.
32. Guo, X.Q.; Wu, W.Y.; Gu, H.R. Phase Locked Loop and Synchronization Methods for Grid- Interfaced Converters: A Review. *Prz. Elektrotechniczny* **2011**, *87*, 182–187.
33. Salamah, A.M.; Finney, S.J.; Willians, B.W. Three-phase Phase-locked loop for Distorted Utilities. *IET Electr. Power Appl.* **2001**, *67*, 263–270.
34. Wiechmann, E.P.; Aqueveque, P.; Burgos, R.; Rodriguez, J. On the Efficiency of Voltage Source and Current Source Inverters for High-Power Drives. *IEEE Trans. Ind. Electron.* **2008**, *55*, 1771–1782. [CrossRef]
35. Wu, B. *High-Power Converters and AC Drives*, 2nd ed.; Wiley: Hoboken, NJ, USA, 2006; pp. 251–316.
36. Al-Falahi, M.D.A.; Tarasiuk, T.; Jayasinghe, S.G.; Jin, Z.; Enshaei, H.; Guerrero, J.M. Ac Ship Microgrids: Control and Power Management Optimization. *Energies* **2018**, *11*, 1458. [CrossRef]
37. Zahedi, B.; Norum, L.E. Modeling and Simulation of All-Electric Ships with Low-Voltage DC Hybrid Power Systems. *IEEE Trans. Power Electron.* **2013**, *28*, 4525–4537. [CrossRef]
38. Thantirige, K.; Rathore, A.K.; Panda, S.K.; Jayasignhe, G.; Zagrodnik, M.A.; Gupta, A.K. Medium Voltage Multilevel Converters for Ship Electric Propulsion Drives. In Proceedings of the 2015 International Conference on Electrical Systems for Aircraft, Railway, Ship Propulsion and Road Vehicles (ESARS), Archen, Germany, 3–5 March 2015.
39. Sofras, E.; Prousalidis, J. Developing a New Methodology for Evaluating Diesel—Electric Propulsion. *J. Mar. Eng. Technol.* **2014**, *13*, 63–92. [CrossRef]
40. ABB GROUP. Available online: http://pdf.nauticexpo.com/pdf/abb-marine/abb-lngcarrier-brochure-final/30709-44457.html (accessed on 1 May 2009).
41. Prousalidis, J.; Antonopoulos, G.; Mouzakis, P.; Sofras, E. On Resolving Reactive Power Problems in Ship Electrical Energy Systems. *J. Mar. Eng. Technol.* **2015**, *14*, 124–136. [CrossRef]
42. Kumar, D.; Zare, F.; Ghosh, A. DC Microgrid Technology: System Architectures, AC Grid Interfaces, Grounding Schemes, Power Quality, Communication Networks, Applications, and Standardizations Aspects. *IEEE Access* **2017**, *5*, 12230–12256. [CrossRef]
43. Sul, S.K. *Control of Electric Machine Drives Systems*; Hongleung Science: Seoul, Korea, 2016; pp. 306–326.

44. ABB. ABB HiPak IGBT Module 5SNA 1200G450350 Datasheet. Available online: https://new.abb.com/semiconductors/integrated-gate-commutated-thyristors- (accessed on 8 January 2019).
45. IEEE. *IEEE Std 519-2014 IEEE Recommended Practice and Requirements for Harmonic Control in Electric Power Systems*; IEEE: New York, NY, USA, 2014; pp. 5–9.

Article

Fuel Consumption and CO_2 Emission Reductions of Ships Powered by a Fuel-Cell-Based Hybrid Power Source

Gilltae Roh [1], Hansung Kim [1], Hyeonmin Jeon [2] and Kyoungkuk Yoon [3,*]

1 Department of Chemical and Biomolecular Engineering, Yonsei University, 134 Shinchon-dong, Seodaemun-gu, Seoul 120-749, Korea

2 Department of Marine System Engineering, Korea Maritime and Ocean University, 727 Taejong-ro, Busan 49112, Korea

3 Department of Electrical Engineering, Colleges of Korea Polytechnic, 155 Sanjeon-ro, Ulsan 44482, Korea

* Correspondence: kkyoon70@kopo.ac.kr; Tel.: +82-(0)1055410424

Received: 17 May 2019; Accepted: 15 July 2019; Published: 18 July 2019

Abstract: The need for technological development to reduce the impact of air pollution caused by ships has been strongly emphasized by many authorities, including the International Maritime Organization (IMO). This has encouraged research to develop an electric propulsion system using hydrogen fuel with the aim of reducing emissions from ships. This paper describes the test bed we constructed to compare our electric propulsion system with existing power sources. Our system uses hybrid power and a diesel engine generator with a combined capacity of 180 kW. To utilize scale-down methodology, the linear interpolation method is applied. The proposed hybrid power source consists of a molten carbonate fuel cell (MCFC), a battery, and a diesel generator, the capacities of which are 100 kW, 30 Kw, and 50 kW, respectively. The experiments we conducted on the test bed were based on the outcome of an analysis of the electrical power consumed in each operating mode considering different types of merchant ships employed in practice. The output, fuel consumption, and CO_2 emission reduction rates of the hybrid and conventional power sources were compared based on the load scenarios created for each type of ship. The CO_2 emissions of the hybrid system was compared with the case of the diesel generator alone operation for each load scenario, with an average of 70%~74%. This analysis confirmed the effectiveness of using a ship with a fuel-cell-based hybrid power source.

Keywords: hybrid power source; fuel cell; molten carbonate fuel cell (MCFC); carbon dioxide; electric propulsion system

---

1. Introduction

In 2015, the "Third IMO Greenhouse Gas Study," conducted by the International Maritime Organization (IMO) [1], reported that air pollutants emitted from ships in 2012 accounted for 13% of NOx, 12% of SOx, and 2.6% of CO_2 in terms of global atmospheric pollutant emissions [2,3]. The International Council on Clean Transport (ICCT), an international environmental non-profit organization, has analyzed and forecasted the pollutant emissions from ships from 1990 to 2050, and reported that the NOx and SOx emitted from ships are expected to increase to 30% and 20%, respectively, of all global pollutant emissions [3–5]. These study results support the view that the long-term effects of atmospheric pollutants caused by ships are foreseen to become more severe, considering the trend of increasing global trade in the future. Clearly, there is a need to develop technology for reducing pollutant emissions from ships [6,7].

Currently, fuel cells that use hydrogen fuel to reduce pollutant emissions from ships are being studied [8–11]. Fuel cells take the chemical energy within the hydrogen that is used as fuel and convert it into electrical and thermal energy through an electrochemical reaction with the oxygen present in the air. They produce almost no pollutant emissions or noise when generating electricity, and they can use various fuels as sources of hydrogen. Fuel cells are an eco-friendly energy source with very high electrical efficiency. Fuel-cell-based power generation can reduce greenhouse gas emissions by 30% compared to existing power generation methods [12]. Therefore, most advanced countries throughout the world regard fuel cells as a next-generation technology and are actively developing them [13,14].

The research on a hybrid system combining a fuel cell system with a diesel engine, which is the main power source of a ship, has been conducted in Europe [15–18]. The report of the European Maritime Safety Agency (EMSA), 'Study on the Use of Fuel Cells in Shipping', applied fuel cells as the main power or auxiliary power of ships from the beginning of 2000 and 24 projects in Europe and the United States with the beginning of the 'US Ship Service Fuel Cell Program [US SSFC]' project [19]. Analysis of these studies shows that most of the methods are generally aimed at improving the performance of hybrid systems (fuel cells, diesel generators) or the configuration of their systems and that there is no experimental study on the reduction of CO_2 emissions from the hybrid power generation systems [20,21].

Therefore, an empirical study was conducted through experiments on CO_2 reduction that has not been carried out in previous projects so far. Molten carbonate fuel cells (MCFCs) were selected as fuel cell systems of a combined power source because the characteristics of MCFCs are suitable for application to ships [22,23]. Since MCFCs operate at high temperature, the reaction rate is fast, even when using low-cost catalysts, as compared with relatively different fuel cell systems. Even when the ship is sailing for a long time, the external reformer is not installed separately and natural gas or coal gas is directly used as fuel. It is appropriate to apply it as the main power source for the base load of the ship [24].

In this study, to reduce the emissions from ships, empirical experiments on the fuel consumption and carbon dioxide emission reduction effect of a combined power source (fuel cell + battery + diesel generator) instead of the diesel generator were conducted. The capacity of the combined power source was 100 kW for MCFCs, 30 kW for batteries, and 50 kW for diesel generators. In order to carry out the experiment on the test bed, the power amount for each operation mode was analyzed according to the type of the commercial vessel, and the scale was downsized according to the capacity of the test bed. The fuel consumption and carbon dioxide emissions of the ship were calculated, according to the load profile of the ship, within 180 kW of the configured system. It can be confirmed through the demonstration that carbon dioxide emission and fuel consumption was considerably reduced compared to the conventional diesel power source.

2. Background

The International Convention for the Prevention of Marine Pollution from Ships (MARPOL) agreed that emissions, such as NOx and SOx, from ships should be reduced by 20% or less of the current emission amounts from 2008 to 2015. Since 2016, the agreement has recommended an 80% reduction in pollutant emissions [25]. In addition, the IMO has introduced the energy efficiency design index (EEDI), which is an index of factors to be considered in ship design to contribute toward reducing CO_2 emissions. The CO_2 emission regulations based on the EEDI that were imposed by the IMO on all new ships built since 2013 are listed in Tables 1 and 2. Ships that do not meet the required EEDI levels are prohibited from entering ports [26]. In Table 2, the EEDI will be implemented in phases. Currently, it is in phase 1, which runs from year 2015 to 2019. Phase 2 will run from year 2020 to 2024 and phase 3 from year 2025 onwards [27].

Table 1. International Convention for the Prevention of Marine Pollution from Ships (MARPOL) 73/78—Annex VI Regulations for the Prevention of Air Pollution from Ships.

Year Built	Capacity	NOx	SOx	PM	Remarks
2008~2015	>125 kW	7.7 g/kWh	24 kg/ton	1.2 kg/ton	20% decrease
2016~		2.0 g/kWh	6 kg/ton	0.3 kg/ton	80% decrease

Table 2. The energy efficiency design index (EEDI)-based CO_2 emission reduction goals.

Phase 0 (Year Built)	Phase 1	Phase 2	Phase 3
2013~2014 Scheduled to take effect	2015~2019 10% decrease	2020~2024 20% decrease	2025~ 30% decrease

National and international regulations are gradually being strengthened and require ocean pollutant emissions from ships to be reduced continuously. However, it is not possible to address this problem solely through modern engine technology without installing additional devices for preventing environmental pollution. Therefore, there is an increasing demand for high-efficiency power sources for ships with almost no pollutant emissions. Normal high-efficiency diesel engines have an energy efficiency of approximately 40%, and facilities equipped with CO_2-capturing devices or pollutant-processing devices for emission gases have limited effectiveness owing to increases in the system volume and fuel energy consumption [28].

On the other hand, if fuel cells powered by hydrogen, which are eco-friendly high-efficiency power sources, could be an alternative solution, instead of diesel engines, to a propel ship, there could be almost no emissions (for example, NOx, SOx, CO_2, or PM); the fuel cells would produce no noise or vibration and would have good power generation efficiency [14]. As such, fuel cells powered by hydrogen have considerable potential as a next-generation main power source for ships. In addition, they can be modularized to reduce complexities in terms of their construction and installation. Therefore, their capacity can be adjusted such that it is most effective for specific types or functions of ships. They have a very wide range of uses and are considered a technology that will play a leading role in ship propulsion systems in the future [8,29].

3. Types and Properties of Fuel Cell Systems

Most fuel cells generate electricity and heat via the chemical reaction between hydrogen and oxygen, and water is created as the product. Various types of fuel cells are being studied, and each of these fuel cells is classified according to the characteristics of its electrolyte. The properties of these cells are described in Figure 1 and Table 3 [19,24].

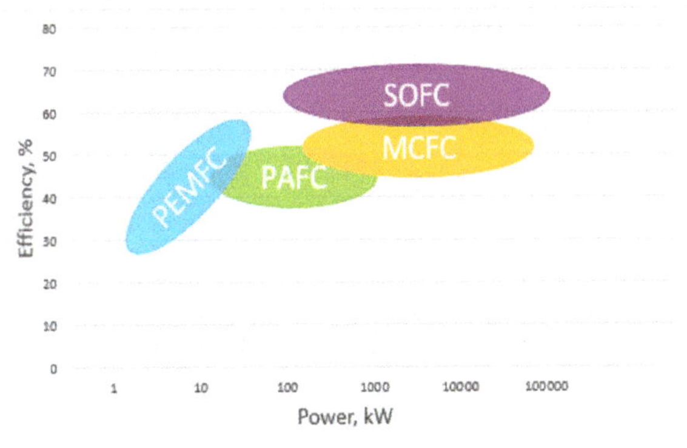

Figure 1. Comparison of efficiency versus power generation in each type of fuel cell.

Table 3. Fuel cell types (catalyst, durability, and hydrogen storage).

Fuel Cell Types	Main Fuel	Electrolyte	Temperature of Generation	Level of Technology	Subject of Application
PEMFC	Hydrogen Methanol	Ion conductive polymer film	Ordinary temperature~100 °C	Development and demonstration phase	Small power source, Automobile
MCFC	Natural gas Coal gas	Molten carbonate	600~700 °C	Development phase	Hybrid power generation
PAFC	Natural gas Methanol	Phosphoric acid	150~200 °C	Commercialization phase	Distributed power system
SOFC	Natural gas Coal gas	Solid oxide	700~1000 °C	Development phase	Hybrid power generation
AFC	Hydrogen	Potassium hydroxide	80~120 °C	Commercialization phase	Space missions

Although there are various types of fuel cells, such as in Table 3, in this paper, MCFCs are considered to be the main power source for the base load of the ship and were applied on the test bed. Because MCFCs operate at a high temperature, they can achieve a fast reaction rate even with a comparatively low-cost catalyst, a simple system design of a fuel cell, and low initial investment cost. In addition, even when the ship is sailing for a long time, natural gas or coal gas can be directly used as a fuel without installing an external reformer separately [16,22,23].

4. Comparative Analysis of Fuel Consumption and CO_2 Emission Reductions in Hybrid Power Sources vs. Conventional Commercial Diesel Generators

4.1. Greenhouse Gas Calculation Method

4.1.1. Intergovernmental Panel on Climate Change (IPCC) Emission Coefficient

Emissions from ships include greenhouse gases (GHGs) emitted from the diesel engines, steam engines with boilers, and gas turbines, which are the main engine types used to power ships, ranging from leisure crafts to large-scale freighters. The emitting crafts, which are the focus of the ship section of the "Intergovernmental Panel on Climate Change (IPCC) Guidelines 2006" report, include sailing ships, fishing boats, and other ships. The method for calculating GHG emissions is presented in Table 4 [30].

Table 4. Calculation methods according to emission gas.

	CO_2	CH_4	N_2O
Estimate methodology	Tier 1,2	Tier 1,2	Tier 1,2

The activity data used in the Tier 1 method are based on fuel consumption, thus emission coefficients are needed for each fuel and pollutant. In the case of CO_2, SO_2, and heavy metals, there is a close relationship between the emission coefficient and the CO_2, SO_2, and heavy metal content of the fuels. The calculations must take into account the related pollutant content in the fuel for each year and the target class of the ship according to the national region.

Tier 1 and Tier 2 emissions were calculated by using a method that uses petroleum sales as the index for the basic level of activity. It performs calculations by assuming the averaged characteristics of each ship type. The method to calculate Tier 3 emissions was based on the operating profile information of the ship. The Tier 3 method can be used when it is possible to collect not only the data on the engine, fuel usage, and duty cycle of the ship, but also information about its voyage. Because the actual voyage data of the ship must be taken into account, port arrival/departure statistical data regarding the voyage of the ship was used to calculate the fuel consumption and emissions while taking into account the emissions for each operating profile and the ship type, fuel type, engine type, technical specifications, and engine load, yearly operating time, etc.

4.1.2. IMO Conversion Emission Factor

At the 1997 MARPOL conference, research on the GHGs emitted by ships was presented via a discussion on "CO_2 emissions from ships." The first GHG study performed by the IMO was presented at the 45th the Marine Environment Protection Committee [MEPC] conference. At the 56th MPEC conference, it was determined that a second IMO GHG study would be performed to examine atmospheric emissions caused by exhaust gas emissions, volatile fuel emissions, and refrigerant leaks.

One goal of this study is to calculate the CO_2 emissions occurring when a hybrid power source is used in a ship. For this, only the exhaust gas emissions of the diesel engine and fuel cell were considered. Although the IPCC calculates GHG emissions by taking into account the ship type, fuel type, engine type, etc., this study used the CO_2 mass conversion factor, a dimensionless constant, presented in the "Calculation of Energy Efficiency Operational Indicator Based on Operational Data" to calculate the CO_2 emissions (IMO MEPC1/Circ 684 2009). GHG emissions were calculated by using the IPCC 2006 guidelines for CH_4 and N_2O in Table 5 (IPCC 2006), and the ISO 8217 Grades DMX conversion factor was used for CO_2 [31–33].

Table 5. Fuel-based exhaust gas emission factors.

Emission	Emission Factor	Guideline Reference
CH_4	0.3 $[TCH_4/TFuel]$	IPCC 2006
N_2O	0.08 $[TN_2O/TFuel]$	IPCC 2006
CO_2	3.206 $[TCO_2/TFuel]$	ISO 8217 Grades DMX

4.2. *Specifications of Components in the Fuel-Cell-Based Hybrid Power Source Test Bed*

The process flow diagram (PFD) of the fuel-cell-based hybrid power source test bed is shown in Figure 2. The test bed was composed of the following specific components: The MCFC system, energy storage system (ESS), diesel generator, load bank, and intelligent energy management system [34–37].

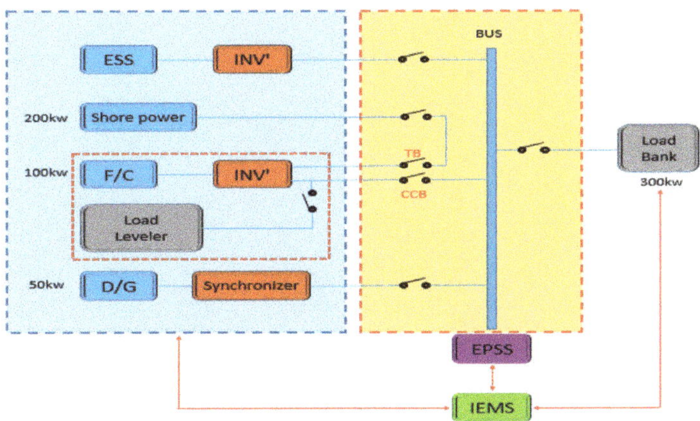

Figure 2. Process flow diagram of the test bed.

4.2.1. MCFC System

The fuel cell used in the test bed was a 300 kW MCFC system composed of a stack module, an electric balance of plant (EBOP), and a machinery balance of plant (MBOP) [38]. The fuel cell system constituting the combined power source was operated with a rated capacity of 300 kW. However, a 100 kW output was used in a practical test bed. The MBOP was pretreated to make a better chemical reaction between the fuel gas and air, which concludes a pre-former, heater, humidifier, valves, pump, and blower [39]. The specifications are listed in Table 6.

Table 6. DFC300 MA system specifications.

DFC300 MA Generation Plant Specifications	
Machinery Balance of Plant [MBOP]	
Height (Main enclosure)	9.6′
Height (Ship loose items)	14.6′
Width	8.0′
Length	19.8′
Weight	12,292 kg
Electric Balance of Plant [EBOP]	
Height	9.5′
Width	3.5′
Length	9.0′
Weight	12,292 kg
Stack Module	8.4′
Height	8.2′
Width	15.0′
Length	18,143 kg
Weight	
Total Weight	42,727 kg
Power output	
Rated output	250 kW
Voltage	380~480 VAC
Frequency	50~60 Hz
Power quality	Per IEEE 519

The peripheral equipment needed by the fuel cell system is shown in Figure 3. It includes a fuel injection part for supplying natural gas, a potable water injection part for producing ultrapure water, a part for emitting drainage water resulting from the production of ultrapure water, and a nitrogen/mixed gas injection part for protecting the stack. The air injection and exhaust gas emission parts were at the top of the MBOP. Two exhaust fans were installed within the MBOP [40].

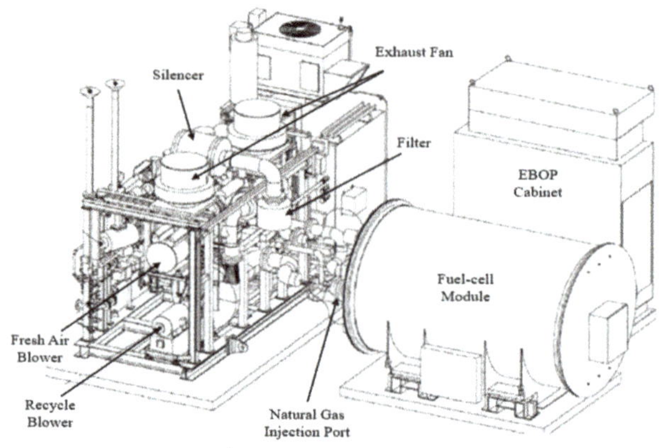

Figure 3. Configuration diagram of peripheral equipment for DFC300 MA.

The power generation concept of the fuel cell system is shown in Figure 4. The system was composed of a heat-up operating mode, which increases the initial temperature of the fuel cell stack module, a ramp-up operating mode, which increases the power to the rated output for actual power generation, and the operation mode, which produces the rated output.

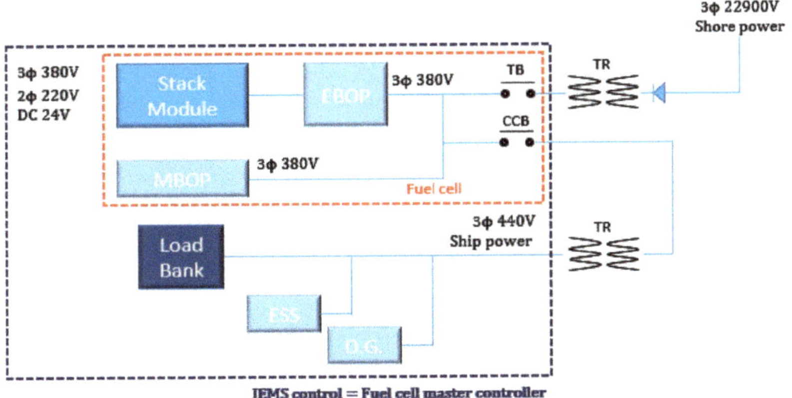

Figure 4. Concept of Power generation for fuel cell system.

4.2.2. Energy Storage System (ESS)

The energy storage system is the electricity storage device, which uses electricity in the battery generated by the fuel cell stored. As shown in Figure 5, it is composed of a secondary battery and power conditioning system (PCS) [41].

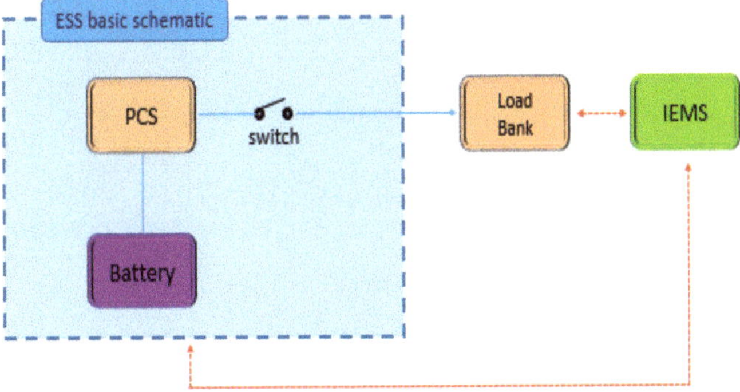

Figure 5. Basic diagram for the energy storage system (ESS).

A lead-acid battery was used for the ESS in the test bed, and it was built using the bidirectional connection system, of which the specifications are listed in Table 7. The PCS has functions for checking the state of charging (SOC) of batteries in real time and controlling the temperature, current, and voltage to enable the system to be operated in a stable manner. It also has functions for surge protection, automated prevention of overcharging/overload, overvoltage alarms, and overvoltage prevention.

Table 7. ESS general battery and inverter specifications.

	Division	Item	Specification
Battery	General	Rated output	100 kW
		Rated voltage	407 VDC
		Capacity	300 AH
		Assembly of Cell	110S/2P
		Range of voltage	352 V~451 V
		Max discharge current	600 A
		Protection	OVP, UVP, OCP, OTP
		Communication	RS 232C, CAN 2.0
Inverter	Alternating Current [AC] Input	Composition	3 Phase
		Voltage	440 VAC
		Frequency	50/60 Hz
		Rated output	100 kW
		Total Harmonic Distortion [THD]	Below 5%
	Output	Voltage	440 VAC
		Frequency	50/60 Hz
		Rated output	100 kW
	Function	Protection	Over current, Over temperature, Over voltage, Low battery shutdown, Reverse flow

4.2.3. Diesel Generator System (DGS)

The 50 kW synchronous generator used in the test bed is a revolving-field-type generator which uses a permanent magnet. Its specifications are given in Table 8.

Table 8. Hybrid test bed and generator specifications.

	Item	Specification
Engine part	Standby power rating	>95 PS
	Engine type	4 Stroke, Water cooled
	Revolution	1800 rpm
	Number of cylinders	6
	Cylinder type	Vertical series
	Governor type	Speed control type
	Cooling system	Radiator type
	Fuel	Diesel
	Starting system	DC 24 V battery start
Generator part	Type	Revolving field magnet
	Standby power rating	50 kW/62.5 kVA
	Prime power rating	45 kW/56 kVA
	Voltage	440/254 V
	Current	82 A
	Phase and wire	Three phase four wire
	Frequency	60 Hz
	Power factor	0.8 Lag
	Pole	4
	Revolution	1800 rpm

4.2.4. Load Bank

The load bank is a forced air-cooled load bank with a rated capacity of 300 kW. It has high resistivity and experiences little change in resistance due to temperature increases. It uses an iron-chrome type 2 heating wire (FCHW-2). The load bank was used in the test bed to provide the electrical load for testing power sources, such as the generator or the uninterruptable power supply. The specifications of the load bank are listed in Table 9. The resistance of the load bank was connected in parallel to allow the load capacity to be adjusted.

Table 9. Load bank specifications.

Rated Capacity		300 kW			
Rated Voltage		3 Phase, 440 V, 60 Hz			
Current and Resistance Value for Each Capacity					
Unit Capacity (kW)	**Quantity**	**Synthetic Capacity (kW)**	**Unit Current (A)**	**Unit Resistance**	**Enclosure Configuration**
0.1	1	0.1	0.13	1936 Ω	1 box
0.2	1	0.2	0.26	968 Ω	
0.4	1	0.4	0.52	484 Ω	
0.8	1	0.8	1.04	242 Ω	
1	1	1	1.31	193.6 Ω	
2	1	2	2.62	96.8 Ω	
4	1	4	5.24	48.4 Ω	
8	1	8	10.49	24.2 Ω	
16	1	16	20.99	12.1 Ω	
32	1	32	41.98	6.05 Ω	
60	4	240	78.78	3.22 Ω	
Total	15	304.5	399.7	-	

4.2.5. Intelligent Energy Management System (IEMS)

The intelligent energy management system (IEMS) is a control system which monitors the voltage, current, and output of each device and the system state in real time and enables the system to be operated reliably. It adjusts the load of the load bank according to the load pattern in real time and allows the different devices to be synchronized [42–46].

The test bed was organized such that the IEMS and power sources (MCFC, diesel generator system (DGS), and ESS) could send and receive device statuses and operation commands through an interface, as shown in Figure 6. Communications were based on an RS-285 and Ethernet to take into account noise and effects of external factors, such as surrounding devices. In addition, an external connection to the internet was used to allow the operating test bed system to be monitored from locations with internet connectivity.

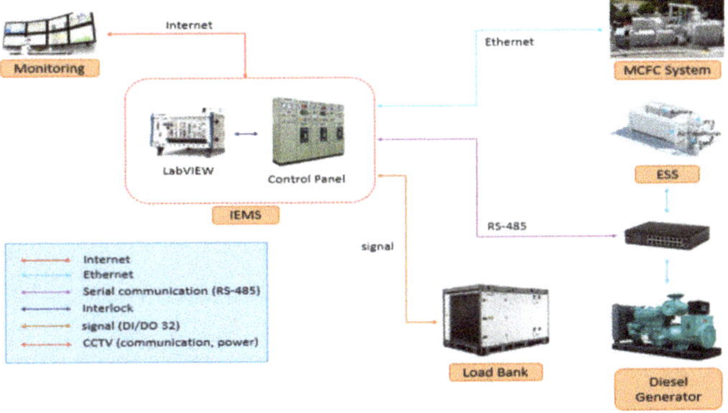

Figure 6. Diagram of interface design.

As shown in the Table 10 below, the control logic was configured to control the complex power system according to the SOC of the ESS according to the load.

Table 10. Configuration of control logic for power system.

Load		Control Condition
≤100 kW	ESS SOC ≥ 80%	MCFC and Consumption Load leveler
	ESS SOC ≤ 80%	MCFC and Charging ESS
101 kW ≤ Load ≤ 130 kW	ESS SOC ≥ 60%	MCFC and Charging ESS and Diesel Generator [D/G] Off
	40% ≤ ESS SOC ≥ 60%	MCFC and ESS Discharging
	ESS SOC ≤ 40%	MCFC and ESS Discharging and D/G Operation
131 kW ≤ Load ≤ 180 kW	ESS SOC ≤ 35%	MCFC and D/G Operation, ESS limit (≤20 kW)
	ESS SOC ≤ 30%	MCFC and D/G Operation, ESS limit (≤10 kW)
	ESS SOC ≤ 25%	MCFC and D/G Operation, ESS off

4.3. Comparison of Fuel Consumption and CO_2 Emission Reduction Rates of the Conventional Commercial Diesel Engine vs. the Hybrid Power Source

4.3.1. Conventional Commercial Diesel Power Source vs. the Fuel-Cell-Based Hybrid Power Source (FCHPS)

The conventional commercial diesel power source was selected from the Doosan Infracore's P126-TI model with a capacity of 241 kW, which was optimized for an average load of 80%. This generator's specific rated power was 80% load of 241 kW, 192.8 kW and selected as the standard model of the diesel generator of the test bed. The diesel generator for fuel cell-based hybrid power source was selected as a DB-58 model with a capacity of 70 kW among the diesel engines of the Doosan Infracore's generator. The generator was also optimized for an average load of 80%, 56 kW was selected as the diesel generator reference model for the combined power source of the test bed.

In order to analyze the CO_2 emissions reduction of the combined power source, the fuel consumption and the carbon dioxide emissions of the commercial diesel generator optimized for the same power as the hybrid power source were applied to the baseload. To compare the fuel consumption of the MCFC and the diesel generator of the combined power source, each fuel consumption amount was converted into the petroleum conversion factor (1 Tonnage of oil equivalent [TOE] = 1000 kgoe). As shown in the Tables 11 and 12, the fuel consumption factor of the diesel generator and the MCFC were matched by applying the energy calorific value conversion factor to each fuel consumption amount for application of the petroleum conversion factor.

Table 11. Flow conversion formula and CO_2 emission calculation formula.

Power Source	Calculation Method of Fuel Consumption	Calculation Method of CO_2
Diesel generator	Kgoe/h = flow (L/h) × 0.901	CO_2 = 0.857 × flow (L/h) × (3.206 + 0.3 + 0.08)
Fuel cell	Kgoe/h = flow (m^3/h) × 1.043	CO_2 = 0.631 × Electric Energy (kWh)

Table 12. Conversion standard of energy calorific value.

Fuel	Unit	Gloss Calorific Value			Net Calorific Value		
		MJ	kcal	10^{-3} TOE	MJ	kcal	10^{-3} TOE
Diesel	L	37.7	9010	0.901	35.2	8420	0.842
LNG	Nm^3	43.6	10,430	1.043	39.4	9420	0.942

Estimation of greenhouse gas emissions was based on the methodology presented in the IPCC Guidelines. In international organizations and countries, emission factors are calculated and used according to the IPCC Guidelines. IMO has also developed a method for estimating carbon dioxide emissions for ships based on the IPCC Guidelines. However, the IPCC Guidelines provide a methodology for estimating emission factors, but IMO suggests a calculation method using conversion factors [47]. Tables 11 and 12 show a calculation formula and conversion standard.

4.3.2. Analysis of Fuel Consumption and CO_2 Emissions Reduction in Hybrid Power Source

The test bed used in this study consisted of a hybrid power source with a combined capacity of 180 kW (100 kW fuel cell, 30 kW battery, and 50 kW diesel generator). The power generation in the hybrid power source was designed such that the fuel cell was set for base-load operation and the battery and diesel generator operated in sequence. At 100 kW in Figure 7, there is a 1% difference in the fuel consumption of the commercial diesel generator and the fuel cell. At 130 kW, the difference in fuel consumption with the diesel generator increases because the fuel cell and the battery, which does not need fuel supply, are operating. At 180 kW, the fuel cell, battery, and diesel generator were operating, and it can be seen that there was a reduction in the fuel consumption and CO_2 emissions of the hybrid power source compared to the commercial diesel generator. At 100 kW in Figure 8, the CO_2 emissions of the fuel cell are 9% of those of the commercial diesel engine. At 130 kW, at which the fuel cell and battery were operating, the difference in CO_2 emissions compared with the diesel generator increases. At 180 kW, at which the fuel cell, battery, and diesel generator were operating, the CO_2 emissions of the hybrid power source were reduced by 39% compared to that of the commercial diesel generator. Table 13 shows the CO_2 emission reduction rates.

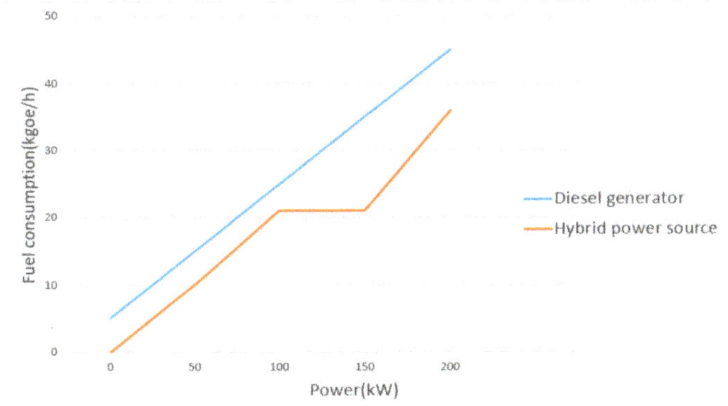

Figure 7. Comparison of fuel consumption of the commercial diesel generator and the hybrid power source.

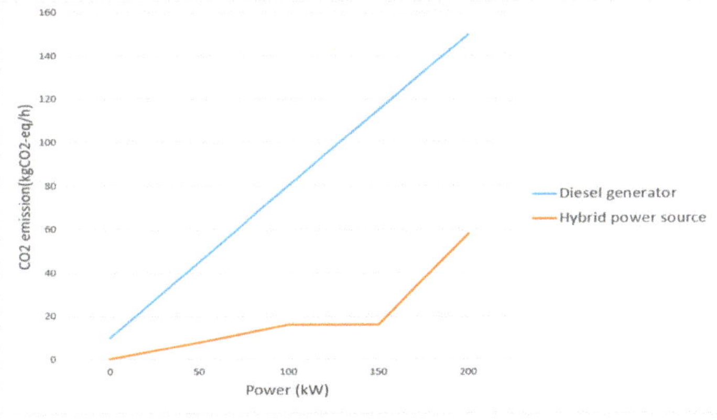

Figure 8. Comparison of CO_2 emissions of the commercial diesel generator and the hybrid power source.

Table 13. CO_2 emission reduction rates in the commercial diesel generator vs. the hybrid power source.

	Fuel Consumption (kgoe/h)	CO_2 Emissions ($kgCO_2/h$)
Diesel generator	43.5	148.5
Hybrid power source	35.6	57.7
CO_2 reduction rate	61%	

5. Analysis of Fuel Consumption and CO_2 Emission Reduction in a Fuel-Cell-Based Hybrid Power Source Using Simulations of Operating Profiles by Type of Ship

The actual electric load analysis values that were used in this study were taken from the operating profiles of ships, including a 5500 TEU Reefer Container, a 13000 TEU Container, a 40 k DWT Bulk Carrier, 130 k DWT LNG Carrier, and 300 k DWT very large crude oil carrier (VLCC). These values were scaled down for each operation mode and suitable load scenarios for each ship type were used. To utilize scale-down methodology, the linear interpolation method is applied [48]. For example, if the original 5500 TEU Reefer Container's rated power is 4154 kW, the rated power of the test bed is 180 kW, when applying the scale down method. At part load, 1424 kW will be converted to 61 kW. All following test bed operating loads were calculated in this way. For the load scenarios in Table 14, according to the ship type operating scenarios, the following power sources were applied.

Table 14. Load scenario according to the ship type.

Vessels	Operation Mode	Power Sources
5500 TEU Reefer Container	Normal seagoing (w/o reefer)	Fuel Cell
	Normal seagoing (w/reefer)	Fuel Cell + Battery + D/G
	Port in/out (w/o thruster)	Fuel Cell
	Port in/out (w/ thruster)	Fuel Cell + Battery + D/G
	Load/Unload	Fuel Cell + Battery + D/G
13,000 TEU Container	Normal seagoing	Fuel Cell
	Port in/out (w/o thruster)	Fuel Cell + Battery
	Port in/out (w/ thruster)	Fuel Cell + Battery + D/G
	Load/Unload	Fuel Cell
	Harboring	Fuel Cell
40 k DWT Bulk Carrier	Normal seagoing	Fuel Cell
	Port in/out	Fuel Cell + Battery + D/G
	Loading (shore crane)	Fuel Cell + Battery
	Loading (crane)	Fuel Cell + Battery + D/G
	Harboring	Fuel Cell
130 k DWT LNG Carrier	Normal seagoing	Fuel Cell
	Port in/out	Fuel Cell + Battery + D/G
	Port discharging	Fuel Cell + Battery + D/G
	Port loading	Fuel Cell + Battery + D/G
	Port idle gas free	Fuel Cell
300 k DWT VLCC	Normal seagoing	Fuel Cell
	W/I.G.S Topping up	Fuel Cell + Battery
	Tank cleaning	Fuel Cell + Battery + D/G
	Port in/out	Fuel Cell + Battery + D/G
	Load/Unload	Fuel Cell + Battery + D/G

5.1. 5500 TEU Reefer Container

The 5500 TEU reefer container uses the following operating modes during operations: Normal seagoing (without reefer), normal seagoing (with reefer), port in/out (without thruster), port in/out (with thruster), and load/unload. To perform the test bed experiments, the scale of the values obtained as a result of the electric load analysis were adjusted to reflect the output of each operating mode of an actual ship. As shown in Figure 9, the hybrid power source was used in the load scenarios of this ship. Normal seagoing (without reefer) was a fuel cell operation interval. Normal seagoing (with reefer) was a fuel cell + battery + diesel generator operation interval. Port in/out (without thruster) was a fuel cell operation interval. Port in/out (with thruster) and load/unload were fuel cell + battery + diesel generator operation intervals. The scale-adjusted electric load analysis was applied to the test bed, and the output tests were carried out.

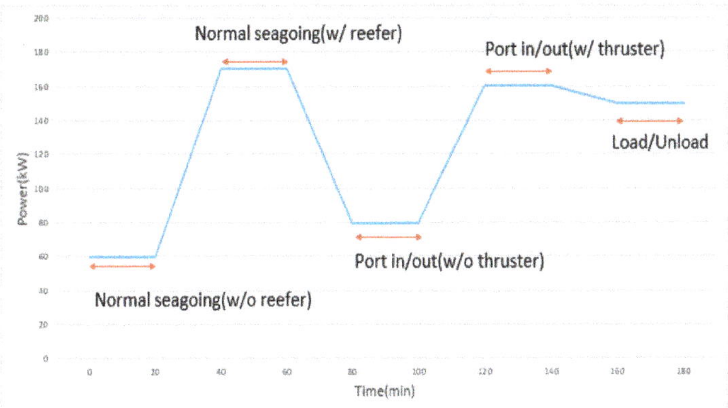

Figure 9. Power consumption of 5500 TEU reefer container during different operation modes.

Figure 10 compares the fuel consumption during each operating mode of this ship. The fuel consumption reached a maximum during the normal seagoing (with reefer) mode and a minimum during the normal seagoing (without reefer) mode. As the load increased, the fuel consumption increased; similarly, as the load decreased, the fuel consumption decreased. However, when observing the CO_2 emission reduction rates shown in Figure 11, it can be seen that the CO_2 emission reduction rate was the highest in the port in/out (without thruster) mode, during which the second least amount of fuel was consumed.

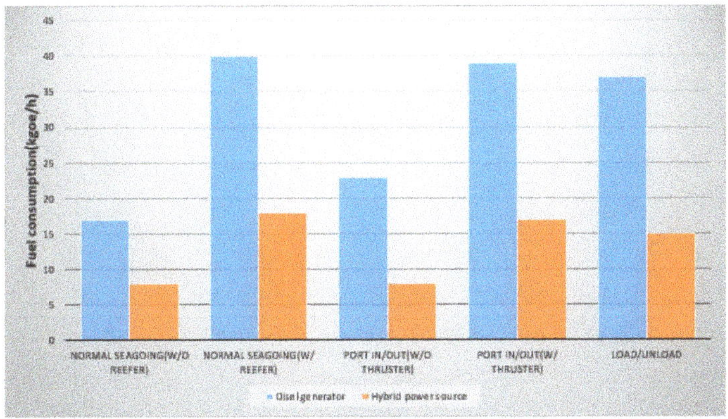

Figure 10. Fuel consumption in each operating mode of the 5500 TEU reefer container.

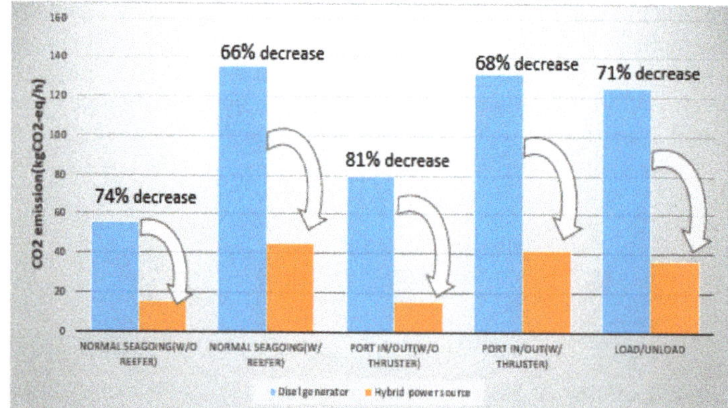

Figure 11. Comparison of CO_2 emissions and CO_2 emission reduction rate in each operating mode of the 5500 TEU reefer container.

5.2. 13000 TEU Container

The 13000 TEU container uses the following operating modes during voyage: Normal seagoing, port in/out (without thruster), port in/out (with thruster), load/unload, and harboring. The test bed experiments were conducted by adjusting the scale of the values obtained from the electric load analysis or the output of each operating mode of an actual ship. As shown in Figure 12, the hybrid power source was used in the load scenarios of this ship. Normal seagoing was a fuel cell operation interval. Port in/out (without thruster) was a fuel cell + battery operation interval. Port in/out (with thruster) was a fuel cell + battery + diesel generator operation interval. Load/unload and harboring were fuel cell operation intervals. The scale-adjusted electric load analysis was applied to the test bed, and the output tests were performed.

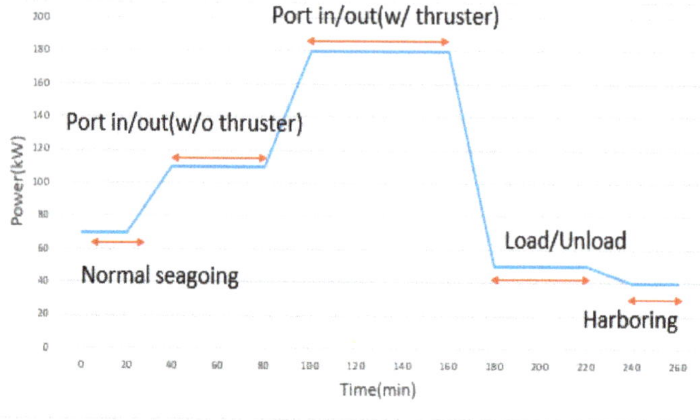

Figure 12. Operating modes of the 13000 TEU container.

Figure 13 compares the fuel consumption during each operating mode of this ship. The fuel consumption was at maximum during the port in/out (with thruster) mode and at minimum during the load/unload mode. The load increased (and decreased) as the fuel consumption increased (and decreased), respectively. However, on observing the CO_2 emission reduction rates shown in Figure 14, it can be seen that the CO_2 emission reduction rate was low in the load/unload and harboring modes despite the low fuel consumption.

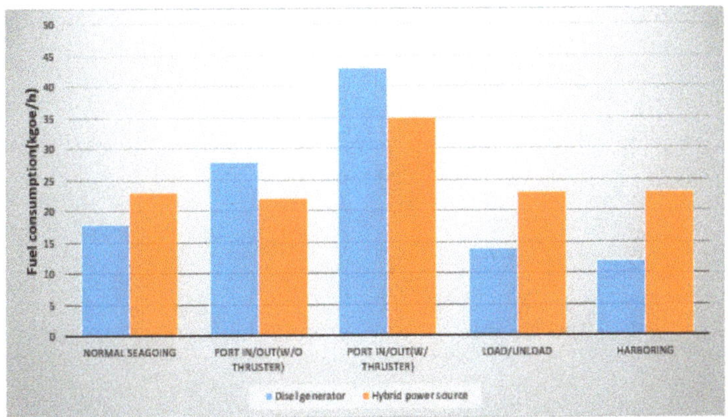

Figure 13. Fuel consumption in each operating mode of the 13000 TEU container ship.

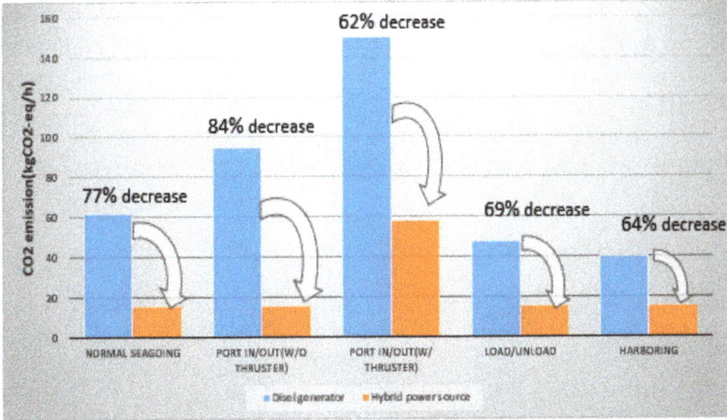

Figure 14. Comparison of CO_2 emissions and CO_2 emission reduction rate in each operating mode of the 13000 TEU container ship.

5.3. 40 k DWT Bulk Carrier

The 40 k DWT bulk carrier uses the following operating modes during operations: Normal seagoing, port in/out, loading (shore crane), loading (deck crane), and harboring. To perform the test bed experiments, the scale of the values obtained as a result of the electric load analysis was adjusted according to the output of each operating mode of an actual ship. As shown in Figure 15, the hybrid power source was used in the load scenarios. Normal seagoing was a fuel cell operation interval. Port in/out was a fuel cell + battery + diesel generator operation interval. Port in/out (with thruster) was a fuel cell + battery + diesel generator operation interval. Loading (shore crane) was a fuel cell + battery operation interval. Loading (deck crane) was a fuel cell + battery + diesel generator operation interval. Harboring was a fuel cell interval. The scale-adjusted electric load analysis was applied to the test bed before the output tests were performed.

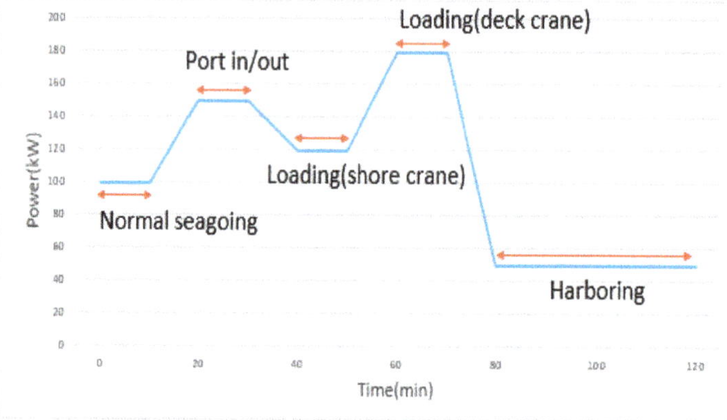

Figure 15. Operating modes of the 40 k DWT bulk carrier.

Figure 16 compares the fuel consumption during each operating mode of this ship. The fuel consumption was at maximum during the loading (deck crane) mode and at minimum during the harboring mode. The fuel consumption increased as the load increased, and decreased as the load decreased. However, on observing the CO_2 emission reduction rates shown in Figure 17, it can be seen that the CO_2 emission reduction rate of the loading (shore crane) mode was as high as 85% even though this operation consumed more fuel than during normal seagoing operations.

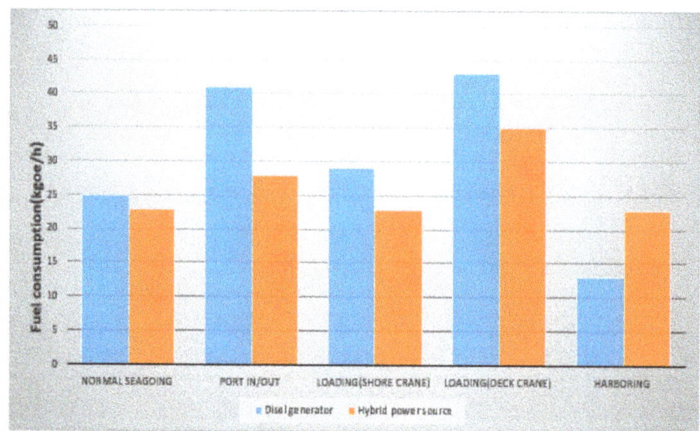

Figure 16. Fuel consumption in each operating mode of the 40 k DWT bulk carrier.

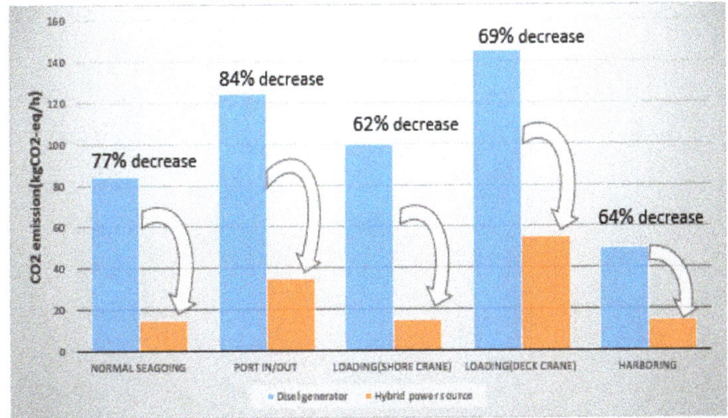

Figure 17. Comparison of CO_2 emissions and CO_2 emission reduction rate in each operating mode of the 40 k DWT bulk carrier.

5.4. 130 k DWT LNG Carrier

The 130 k DWT LNG carrier uses the following operating modes during its operations: Normal seagoing, port in/out, port discharging, port loading, and port idle gas free. The test bed experiments were conducted by adjusting the scale of the values obtained from the electric load analysis based on the output of each operating mode of an actual ship. As shown in Figure 18, the hybrid power source was used in the load scenarios of this ship. Normal seagoing was a fuel cell operation interval. Port in/out, port discharging, and port loading were fuel cell + battery + diesel generator operation intervals. Port idle gas free was a fuel cell operation interval. The scale-adjusted electric load analysis was applied to the test bed, and the output tests were performed.

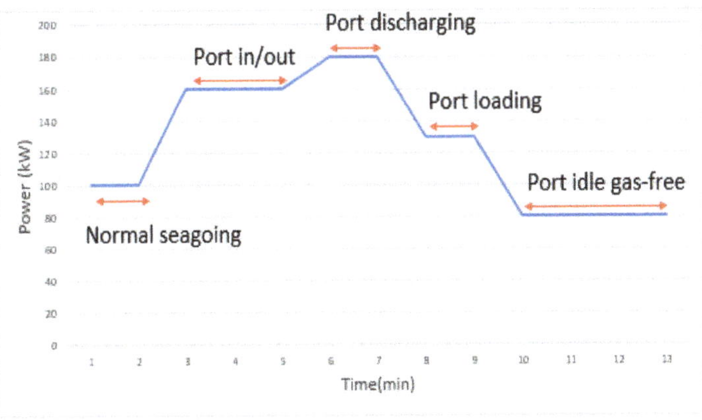

Figure 18. Operating modes of the 130 k DWT LNG carrier.

Figure 19 compares the fuel consumption during each operating mode of this ship. The maximum amount of fuel was consumed during the port discharging mode, whereas it reached a minimum during the port idle gas free mode. The fuel consumption increased and decreased as the load increased and decreased, respectively. However, on examining the CO_2 emission reduction rates shown in Figure 20, it can be seen that the CO_2 emission reduction rate of the normal seagoing mode was as high as 83%, even though the fuel consumption in this mode exceeded that in the port idle gas free mode.

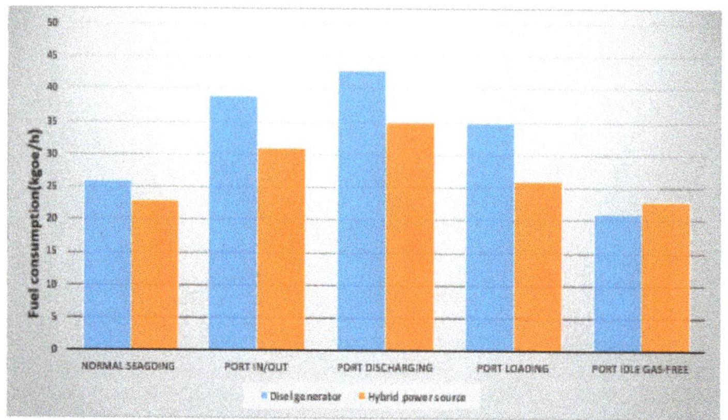

Figure 19. Fuel consumption in each operation mode of the 13 0k DWT LNG carrier.

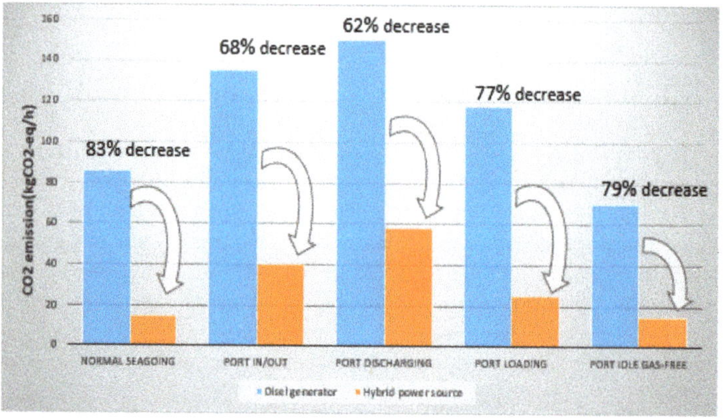

Figure 20. Comparison of CO_2 emissions and CO_2 emission reduction rate in each operating mode of the 130 k DWT LNG carrier.

5.5. 300 k DWT Very Large Crude Oil Carrier (VLCC)

The 300 k DWT VLCC uses the following operating modes during operations: Normal seagoing, with an inert gas supply system (IGS) topping up, tank cleaning, port in/out, and load/unload. An adjustment was made to the scale of the values of the electric load analysis of the output of each operating mode of an actual ship to perform the test bed experiments. As shown in Figure 21, the hybrid power source was used in the load scenarios of this ship. Normal seagoing was a fuel cell operation interval. With IGS topping up was a fuel cell + battery operation interval. Tank cleaning, port in/out, and load/unload were fuel cell + battery + diesel generator operation intervals. The scale-adjusted electric load analysis was applied to the test bed, and the output tests were performed.

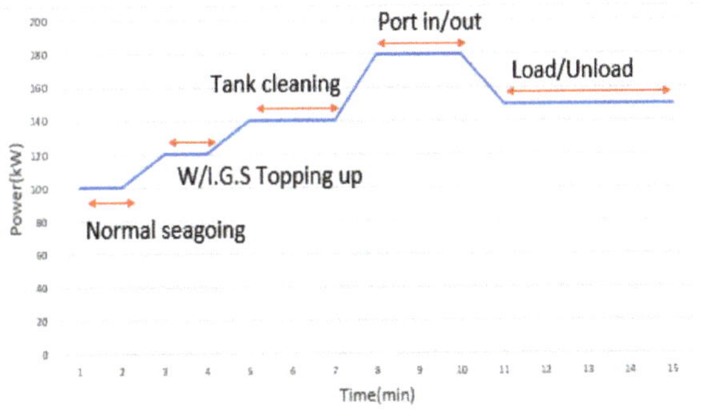

Figure 21. Operating modes of the 300 k DWT VLCC.

Figure 22 compares the fuel consumption during each operating mode of this ship. The fuel consumption was at maximum during the port in/out mode and at minimum during the normal seagoing mode. The fuel consumption increased as the load increased and decreased as the load decreased. However, on observing the CO_2 emission reduction rates shown in Figure 23, it can be seen that the CO_2 emission reduction rate in the IGS topping up mode was as high as 85%, even though the fuel consumption in this mode was higher than in normal seagoing mode.

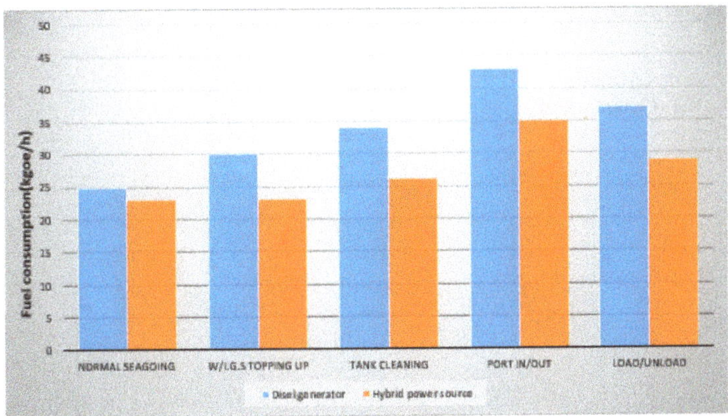

Figure 22. Fuel consumption in each operating mode of the 300 k DWT VLCC.

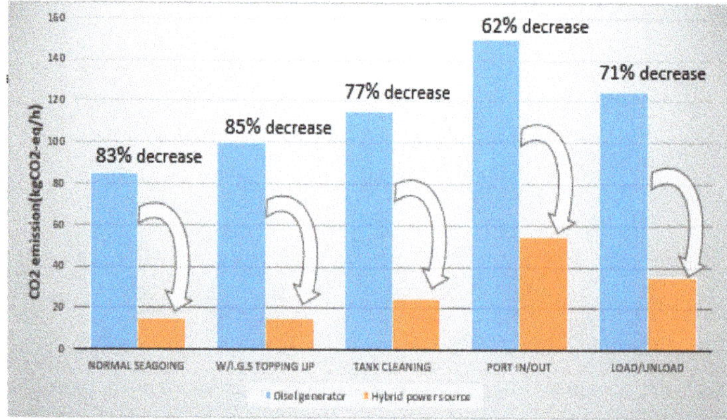

Figure 23. Comparison of CO_2 emissions and CO_2 emission reduction rate in each operating mode of the 300 k DWT VLCC.

Operating profile scenarios for each type of ship were developed, and the five developed load scenarios were applied to the test bed. The results are presented in Table 15.

Table 15. Cumulative CO_2 emissions and reductions at load scenario.

Case	CO_2 Emissions ($kgCO_2$)		Fuel Consumption (kgoe)		CO_2 Reduction Rate	Experiment Time (h:m:s)
	Diesel Generator	Hybrid Power Source	Diesel Generator	Hybrid Power Source		
5500 TEU Reefer Container	205.6	56.8	60.3	54.1	72	2:50:59
13,000 Container	233.9	69.5	68.6	57.8	70	4:01:01
40 k DWT Bulk Carrier	184.9	49.8	54.2	52.0	73	2:01:07
130 k DWT LNG Carrier	217.7	59.6	63.8	54.9	73	2:00:23
300 k DWT VLCC	238.5	62.8	69.9	55.8	74	2:00:20

6. Conclusions

This study analyzed the fuel consumption and CO_2 emission reduction rates when a fuel-cell-based hybrid power source instead of a conventional commercial diesel power source was used in ships. The results showed that under the rated output on a test bed with a load bank of 180 kW, the conventional commercial diesel generator consumed fuel at 43.5 kgoe/h and emitted CO_2 at 148.5 kg/h, whereas the fuel-cell-based hybrid power source consumed fuel at 35.6 kgoe/h and emitted CO_2 at 57.7 kg/h, as given in Table 11. The hybrid power source reduced fuel consumption by 18% and CO_2 emissions by 61% at part load in the port period. These results indicate that it is possible to reduce CO_2 emissions by up to 61% if a hybrid power source of the same capacity is used to power a ship.

In this study, the actual electric load analysis values of the 5500 TEU Reefer Container, 13 k TEU Container, 40 k Bulk Carrier, 130 k DWT LNG Carrier, and 300 k DWT Crude Oil Tanker were scaled down according to the operation mode, and the control logic and systems of the test bed developed in this study were operated normally according to the respective load scenarios. The experimental results of applying the developed five load scenarios to the test bed are shown in Table 15.

Because the output characteristics and control time of the diesel generator, according to the power source of the hybrid system, were reduced, according to the load variation pattern of the ship and the ship's type, the CO_2 emissions of the hybrid system, as compared with the case of the diesel generator, alone operated for each load scenario with an average of 70%~74% less.

In order to apply the hybrid system to ships, it is possible to maximize the CO_2 emission reduction effect by setting the capacity of the fuel cell + battery to be able to take charge of the base load of the ship through analysis of the base load of each ship type.

Author Contributions: Conceptualization, G.R., H.K. and K.Y.; Methodology, G.R., K.Y. and H.J.; Formal analysis, G.R. and H.J.; Supervision, H.K. and K.Y., Visualization, H.J; Writing-original draft preparation, G.R. and K.Y.; Writing-review and editing, K.Y.

Funding: This research received no external funding.

Acknowledgments: We thank our colleagues from Jong Su Kim who provided insight and expertise that greatly assisted the research, although they may not agree with all of the conclusions of this paper.

Conflicts of Interest: The authors declare no conflicts of interest.

Nomenclature

AFC	Alkaline fuel cell
DWT	Deadweight tons
EBOP	Electric balance of plant
EEDI	Energy efficiency design index
EPSS	Electric power switching system
ESS	Energy storage system
IEMS	Intelligent energy management system
IGS	Inert gas system
IMO	International maritime organization
IPCC	Intergovernmental panel on climate change
KGOE	Kilograms of oil equivalent
MARPOL	The international convention for the prevention of marine pollution from ships
MCFC	Molten carbonate fuel cell
MDOP	Machinery balance of plant
PAFC	Phosphoric acid fuel cells
PCS	Power conditioning system
PEMFC	Polymer electrolyte membrane fuel cell
PFD	Process flow diagram
PM	Particulate matter
SFC	Specific fuel consumption
SOC	State of charge
SOFC	Solid oxide fuel cell
TEU	Twenty-foot equivalent units
VLCC	Very large crude-oil carrier

References

1. International Maritime Organization. Third IMO Greenhouse Gas Study. Available online: http://www.imo.org/en/OurWork/Environment/PollutionPrevention/AirPollution/Documents/Third%20Greenhouse%20Gas%20Study/GHG3%20Executive%20Summary.pdf (accessed on 15 August 2015).
2. Bouman, E.A.; Lindstad, E.; Rialland, A.I.; Strømman, A.H.; Lindstad, H.E. State-of-the-art technologies, measures, and potential for reducing GHG emissions from shipping—A review. *Transp. Res. Part D Transp. Environ.* **2017**, 52, 408–421. [CrossRef]

3. European Parliament. Emission Reduction Targets for International Aviation and Shipping. Available online: http://www.europarl.europa.eu/RegData/etudes/STUD/2015/569964/IPOL_STU(2015)569964_EN.pdf (accessed on 18 September 2015).
4. International Council on Clean Transportaion. Greenhouse Gas Emissions from Global Shipping. Available online: https://www.theicct.org/sites/default/files/publications/Global-shipping-GHG-emissions-2013-2015_ICCT-Report_17102017_vF.pdf (accessed on 1 October 2017).
5. International Maritime Organization. Marine Environment Protection Committee 72nd Session. Available online: http://www.imo.org/en/MediaCentre/MeetingSummaries/MEPC/Pages/MEPC-72nd-session.aspx (accessed on 13 April 2018).
6. Pestana, H. Future trends of electrical propulsion and implications to ship design. *Marit. Technol. Eng.* **2014**, *1*, 797–803.
7. Yang, Z.; Zhang, D.; Caglayan, O.; Jenkinson, I.; Bonsall, S.; Wang, J.; Huang, M.; Yan, X.; Yang, Z. Selection of techniques for reducing shipping NOx and SOx emissions. *Transp. Res. Part D Transp. Environ.* **2012**, *17*, 478–486. [CrossRef]
8. Sattler, G. Fuel Cells Going On-Board. *J. Power Sources* **2000**, *86*, 61–67. [CrossRef]
9. Tronstad, T.; Endresen, Ø. Fuel cells for low emission ships. In Proceedings of the International Renewable Energy Conference 2005, Beijing, China, 7–8 November 2005; pp. 141–149.
10. Schneider, J.; Dirk, S.; Schneider, J.; Dirk, S.; Motor, P. ZEMShip. In Proceedings of the 18th World Hydrogen Energy Conference 2010, Essen, Germany, 16–20 May 2010; Volume 78, pp. 132–135.
11. Yuan, J.; Ng, S.H.; Sou, W.S. Uncertainty quantification of CO_2 emission reduction for maritime shipping. *Energy Policy* **2016**, *88*, 113–130. [CrossRef]
12. Van Biert, L.; Godjevac, M.; Visser, K.; Aravind, P. A review of fuel cell systems for maritime applications. *J. Power Sources* **2016**, *327*, 345–364. [CrossRef]
13. De-Troya, J.J.; Álvarez, C.; Fernández-Garrido, C.; Carral, L. Analysing the possibilities of using fuel cells in ships. *Int. J. Hydrog. Energy* **2016**, *41*, 2853–2866. [CrossRef]
14. Andújar, J.; Segura, F.; Andujar-Márquez, J.M. Fuel cells: History and updating. A walk along two centuries. *Renew. Sustain. Energy Rev.* **2009**, *13*, 2309–2322. [CrossRef]
15. Inal, O.B.; Deniz, C. Fuel Cell Availability for Merchant Ships Fuel Cell Availability for Merchant Ships. In Proceedings of the 3rd International Symposium on Naval Architecture and Maritime (INT-NAM 2018), İstanbul, Turkey, 23–25 April 2018; Volume 1, pp. 907–916.
16. Tomczyk, P. MCFC versus other fuel cells—Characteristics, technologies and prospects. *J. Power Sources* **2006**, *160*, 858–862. [CrossRef]
17. Vogler, F.; Würsig, G. New Developments for Maritime Fuel Cell Systems. In Proceedings of the 18th World Hydrogen Energy Conference (WHEC 2010), Essen, Germany, 16–21 May 2010; Volume 78, pp. 444–453.
18. McConnell, V.P. Now, Voyager? The Increasing Marine Use of Fuel Cells. *Fuel Cells Bull.* **2010**, *2010*, 12–17. [CrossRef]
19. Tronstad, T.; Åstrand, H.H.; Haugom, G.P.; Langfeldt, L. *Study on the Use of Fuel Cells in Shipping*; EMSA European Maritime Safety Agency: Lisbon, Portugal, 2017; Volume 1, pp. 1–108.
20. Yazdanfar, J.; Mehrpooya, M.; Yousefi, H.; Palizdar, A. Energy and exergy analysis and optimal design of the hybrid molten carbonate fuel cell power plant and carbon dioxide capturing process. *Energy Convers. Manag.* **2015**, *98*, 15–27. [CrossRef]
21. Milewski, J.; Lewandowski, J.; Miller, A. Reducing CO_2 emissions from coal fired power plant by using a molten carbonate fuel cell. In Proceedings of the ASME Turbo Expo 2008, Berlin, Germany, 9–13 June 2008; pp. 1–7.
22. Welaya, Y.M.; El Gohary, M.M.; Ammar, N.R. A comparison between fuel cells and other alternatives for marine electric power generation. *Int. J. Nav. Arch. Ocean Eng.* **2011**, *3*, 141–149. [CrossRef]
23. Mekhilef, S.; Saidur, R.; Safari, A. Comparative study of different fuel cell technologies. *Renew. Sustain. Energy Rev.* **2012**, *16*, 981–989. [CrossRef]
24. Larminie, J.; Dicks, A. *Fuel Cell Systems Explained*; Wiley: Hoboken, NJ, USA, 2003.
25. International Maritime Organization. MARPOL Convention 73/78 Annex VI. Available online: http://www.imo.org/en/About/Conventions/ListOfConventions/Pages/International-Convention-for-the-Prevention-of-Pollution-from-Ships-(MARPOL).aspx (accessed on 19 May 2005).

26. International Maritime Organization. IMO MEPC1/Circ 684. *Guidelines for Voluntary Use of the Ship Energy Efficiency Operational Indicator (EEOI)*. Available online: https://www.classnk.or.jp/hp/pdf/activities/statutory/eedi/mepc_1-circ_684.pdf (accessed on 17 August 2009).
27. International Maritime Organization. Ship Energy Efficiency Regulations and Related Guidelines. Available online: http://www.imo.org/en/OurWork/Environment/PollutionPrevention/AirPollution/Documents/Air%20pollution/M2%20EE%20regulations%20and%20guidelines%20final.pdf (accessed on 2 January 2016).
28. Woodyard, D. *Pounder's: Marine Diesel Engines and Gas Turbines*; Elsevier: Burlington, UK, 2004.
29. Zahedi, B.; Norum, L.E.; Ludvigsen, K.B. Optimized efficiency of all-electric ships by dc hybrid power systems. *J. Power Sources* **2014**, *255*, 341–354. [CrossRef]
30. Intergovernmental Panel on Climate Change. IPCC Guidelines for National Greenhouse Gas Inventories. Available online: https://www.ipcc-nggip.iges.or.jp/public/2006gl/vol2.html (accessed on 10 January 2006).
31. ISO. ISO 8271:2012 Petroleum Products—Specifications of Marine Fuels. Available online: https://www.iso.org/obp/ui/#iso:std:iso:8217:ed-5:v1:en (accessed on 5 March 2012).
32. International Maritime Organization. Marine Environment Protection Committee 65th Session Agenda Item 4. Available online: http://yasinskiy.net/docs/wp-content/uploads/2018/06/MEPC.-65-INF.3-Rev.1.pdf (accessed on 11 December 2012).
33. Park, S.-K.; Youn, Y.-M. Analysis of International Standardization Trend for the Application of Fuel Cell Systems on Ships. *J. Korean Soc. Mar. Environ. Saf.* **2015**, *20*, 579–585. [CrossRef]
34. Inci, M.; Türksoy, Ö. Review of fuel cells to grid interface: Configurations, technical challenges and trends. *J. Clean. Prod.* **2019**, *213*, 1353–1370. [CrossRef]
35. Kirubakaran, A.; Jain, S.; Nema, R. A review on fuel cell technologies and power electronic interface. *Renew. Sustain. Energy Rev.* **2009**, *13*, 2430–2440. [CrossRef]
36. Park, K.-D.; Roh, G.-T.; Kim, K.-H.; Ahn, J.-W. Experimental study on development of a marine hybrid power system. *J. Korean Soc. Mar. Eng.* **2017**, *41*, 495–500. [CrossRef]
37. Bassam, A.M.; Phillips, A.B.; Turnock, S.R.; Wilson, P.A. An improved energy management strategy for a hybrid fuel cell/battery passenger vessel. *Int. J. Hydrog. Energy* **2016**, *41*, 22453–22464. [CrossRef]
38. Fuel Cell Energy Inc. 300kW DFC300 Specification. Available online: http://www.fuelcellenergy.com/assets/PID000155_FCE_DFC300_r1_hires.pdf (accessed on 16 August 2011).
39. Milewski, J.; Świercz, T.; Badyda, K.; Miller, A.; Dmowski, A.; Biczel, P. The control strategy for a molten carbonate fuel cell hybrid system. *Int. J. Hydrog. Energy* **2010**, *35*, 2997–3000. [CrossRef]
40. Heidebrecht, P.; Sundmacher, K. Molten carbonate fuel cell (MCFC) with internal reforming: Model-based analysis of cell dynamics. *Chem. Eng. Sci.* **2003**, *58*, 1029–1036. [CrossRef]
41. Cao, J.; Emadi, A. A New Battery/Ultracapacitor Hybrid Energy Storage System for Electric, Hybrid, and Plug-in Hybrid Electric Vehicles. *IEEE Trans. Power Electron.* **2012**, *27*, 122–132.
42. Choi, C.H.; Yu, S.; Han, I.-S.; Kho, B.-K.; Kang, D.-G.; Lee, H.Y.; Seo, M.-S.; Kong, J.-W.; Kim, G.; Ahn, J.-W.; et al. Development and demonstration of PEM fuel-cell-battery hybrid system for propulsion of tourist boat. *Int. J. Hydrog. Energy* **2016**, *41*, 3591–3599. [CrossRef]
43. Han, J.; Charpentier, J.-F.; Tang, T. An Energy Management System of a Fuel Cell/Battery Hybrid Boat. *Energies* **2014**, *7*, 2799–2820. [CrossRef]
44. San Martín, J.I.; Zamora, I.; Asensio, F.J.; García Villalobos, J.; San Martín, J.J.; Aperribay, V. Control and Management of a Fuel Cell Microgrid. Energy Efficiency Optimization. *Renew. Energy Power Qual. J.* **2017**, *7*, 314–319. [CrossRef]
45. Garcia, P.; Fernandez, L.M.; Garcia, C.A.; Jurado, F.; Fernández-Ramirez, L.M. Energy Management System of Fuel-Cell-Battery Hybrid Tramway. *IEEE Trans. Ind. Electron.* **2010**, *57*, 4013–4023. [CrossRef]
46. Jabir, M.; Mokhlis, H.; Muhammad, M.A.; Illias, H.A. Optimal battery and fuel cell operation for energy management strategy in MG. *IET Gener. Transm. Distrib.* **2019**, *13*, 997–1004. [CrossRef]

47. National Law Information Center of Korea. Enforcement Rule of Energy Law. Available online: http://law.go.kr/nwRvsLsInfoR.do?lsiSeq=200561 (accessed on 28 December 2017).
48. Ha, Y.-J.; Lee, Y.-G.; Kang, B.H. A Study on the Estimation of the Form Factor of Full-Scale Ship by the Experimental Data of Geosim Models. *J. Soc. Nav. Arch. Korea* **2013**, *50*, 291–297. [CrossRef]

Journal of
Marine Science and Engineering

Article

Fuel Cell Application for Investigating the Quality of Electricity from Ship Hybrid Power Sources

Hyeonmin Jeon [1], Seongwan Kim [1] and Kyoungkuk Yoon [2,*]

1 Department of Marine System Engineering, Korea Maritime and Ocean University, 727 Taejong -ro, Busan 49112, Korea
2 Department of Electrical Engineering, Colleges of Korea Polytechnic, 155 Sanjeon-ro, Ulsan 44482, Korea
* Correspondence: kkyoon70@kopo.ac.kr; Tel.: +82-105-541-0424

Received: 2 July 2019; Accepted: 19 July 2019; Published: 26 July 2019

Abstract: Since recent marine application of fuel cell systems has been due largely limited to small-sized ships, this paper was aimed to investigate the technical applicability of molten carbonate fuel cell (MCFC) for medium and large-sized ships, using a 180 kW class hybrid test bed with combined power sources: A 100 kW MCFC, a 30 kW battery and a 50 kW diesel generator. This study focused primarily on determining whether the combined system designed in consideration of actual marine power system configuration could function properly. A case study was conducted with a 5500 Twenty-foot Equivalent Unit (TEU) container vessel. The operation profile was collected and analyzed in order to develop electric load scenarios applicable to the power system. Throughout the experiment, we evaluated the power quality of the voltage and frequency in the process of synchronization and de-synchronization across the power sources. Therefore, research results revealed that power quality continued to be excellent. This outcome provides insight into the technical reliability of MCFC application on large marine vessels.

Keywords: Molten carbonate fuel cell (MCFC); Hybrid test bed; Operation profile; Power quality

---

1. Introduction

Recently, hydrogen has started to be regarded as an alternative ocean fuel source. As a result, studies on hydrogen-fueled fuel cells have been actively conducted in Europe and the United States [1–3].

Fuel cells are known as a technology in which the chemical energy associated with hydrogen molecules is converted to electricity and thermal energy through electrochemical reactions with air. Thanks to the minimization of emissions, the noise in operation and high adaptability of various fuel sources, fuel cells are considered a next-generation technology for clean production [4,5].

In particular, fuel cell-based power plants are expected to reduce greenhouse gas emissions by 30% compared to fossil fuel-based power generation. Considering such benefits, many developed countries have been carried out a number of projects to stimulate the fuel cell application to industries [4–10].

According to a report of the European Maritime Safety Agency (EMSA), since the first project 'US SSFC' was launched in 2000, 24 projects have been made available to facilitate the application of marine fuel cells. Eleven of them selected the Proton Exchange Membrane Fuel Cell (PEMFC) type composed of polymer resin with several advantages: Relatively low operating temperature, about 80 °C; shorter time to reach the operating temperature; unnecessary of peripheral devices [11].

On the other hand, to ensure the reliability of the stakes, it requires pure hydrogen which is operated in a very low temperature with the late response time. In order to reduce such drawbacks, expensive catalyst and electrode are applied [12–14].

To obtain pure hydrogen, a separate reforming system is additionally required. That limits the application of PEMFC for the propulsion of medium-large ships. The Zero Emissions Ships (ZEMShip)

project has demonstrated the excellent performance of a fuel cell by applying a 96 kW fuel cell to the Motor Vessel (MV) 'Alsterwasser' [15–19].

In 'FellowSHIP' and 'Molten Carbonate Fuel Cells for Waterborne Application (MC-WAP)' projects, molten carbonate fuel cell (MCFC) type was applied to supplement auxiliary power rather than the main propulsion power. In particular, 'FellowSHIP' from 2003 to 2017, applied MCFC fuel cells for MV 'Viking Lady', 6,000 Deadweight Tonnage (DWT) Offshore Support Vessel (OSV), The project results showed excellent performance of the fuel cell in reducing emission levels [8,20–22].

Because MCFCs can operate at high temperatures, low-cost catalysts are available, which simplifying system design and reducing costs. In addition, even with long voyages, this type of fuel cell can utilize natural gas or coal gas as a direct fuel instead of using an external reformer. These advantages may be suitable for application as a major source of power for the ship's fundamental loads. Despite many research and projects for fuel cell applications in the marine industry, attempts to use the MCFC type for medium and large vessels for propulsion are scarce. Given this background, this study was motivated to investigate the suitability of the MCFC for the large vessel propulsion [23–27].

This paper was focused on analyzing the power quality of each power source in synchronization and breakaway. A test bed with a capacity of 180 kW was constructed using 100 kW fuel cells, 30 kW batteries and 50 kW diesel generator.

An actual voyage data for 5,500 TEU container vessel was used to verify the power quality of the fuel cell. Three load scenarios were developed by examining the performance characteristics of the fuel cell, battery and diesel generator systems based on normal navigational conditions. Each developed scenario was applied to the hybrid power sources, and the quality of voltage and frequency was examined during synchronization and breakaway phases.

2. Methodology

Given that the past research was largely focused on the marine application of MCFC as the main power source for small-sized vessel, this research would be a record of the first research for investigating the practicability of MCFC for medium- and large-sized vessels. To achieve this goal, a 5,500 TEU class container ship was selected as a case ship, and its operating data was applied for the test bed simulation in order to analyze the power quality of the system.

For this research, a combined power source test bed composed of MCFC, battery and diesel engine was constructed. Based on the developed load scenario from the case ship, we analyzed the quality of voltage and frequency in response to the synchronization and de-synchronization between each power source in the test bed and evaluated whether the results could meet the current regulations. Figure 1 shows the outline of the research process.

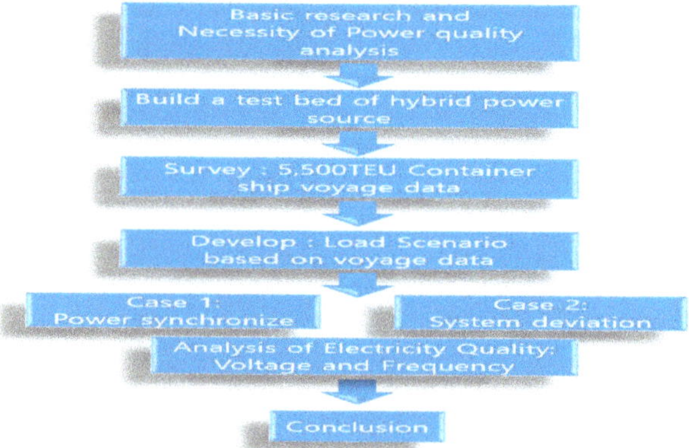

Figure 1. Flowchart for research procedure.

3. Design and Construction of Fuel Cell-Based Hybrid Power Source Test Bed

3.1. Design

Figure 2 shows the basic structure of the energy management system (EMS) used to control the hybrid power system in the test bed in accordance with load variations. In detail, the main parts of the test bed consist of MCFC system, energy storage system (ESS), diesel generator, load bank and EMS and electric power switching system (EPSS). In addition, the test bed was also optimally programmed to reliably synchronize hybrid power systems.

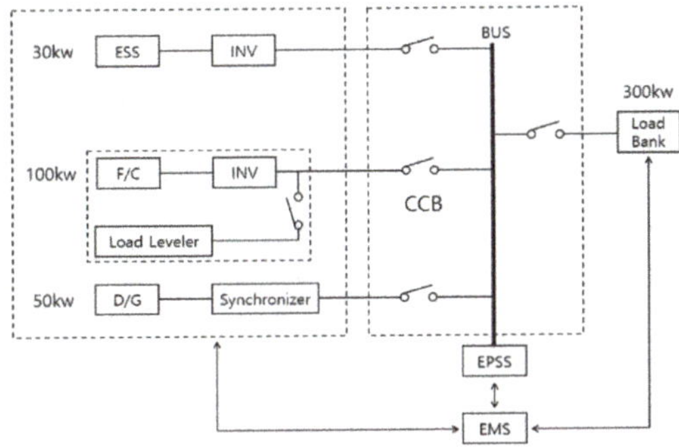

Figure 2. Basic outline of hybrid power systems for the test bed.

3.2. Components of the Test Bed for Fuel Cell-Based Ship Hybrid Power Systems

Figure 3 shows the arrangement of the fuel cell-based ship hybrid power system test bed.

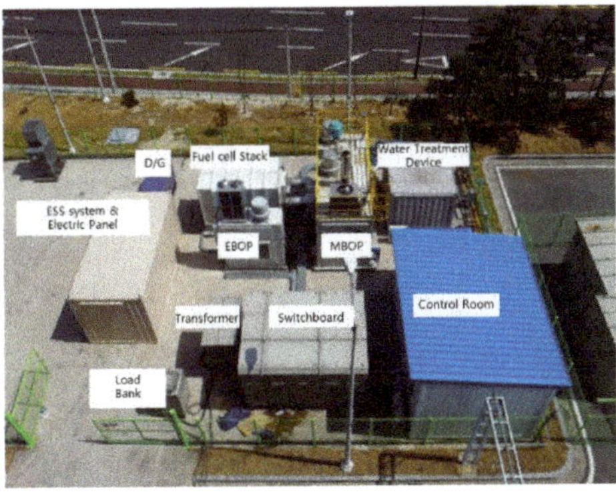

Figure 3. Actual placement of the test bed.

3.2.1. MCFC System

The fuel cell to be applied to the test bed is a 300 kW class MCFC system 'DFC300MA' model manufactured by 'POSCO Engineering' [28]. It is also a model that 'FuelCell Energy (FCE)' in the USA has developed as a basic model. Various development and experiments in several stages made the most optimized system at present. The fuel cell system consists of a stack module. EBOP (electric balance of plant) and MBOP (machinery balance of plant). The specification of the fuel cell system is shown in Table 1.

Table 1. Specification of the fuel cell system.

Specification for 'DFC300MA'.	
Power Output	-
-Rated output	250 kW
-Voltage	380 - 480 VAC
Frequency	50 ~ 60 Hz
-Power Quality	Per IEEE 519
Emissions	-
-NOx	<0.02 lb/MWh
-SOx	<0.001 lb/MWh
-CO	<0.05 lb/MWh

Figure 4 shows the peripheral equipment for the fuel cell system: Particularly, a fuel injection part for natural gas supply, a water injection part for making ultrapure water, a water discharge part and nitrogen/mixed gas injection part for stack protection. The Air injection part and the water discharge part are located on the upper side of MBOP within which two sets of exhaust fans are fitted.

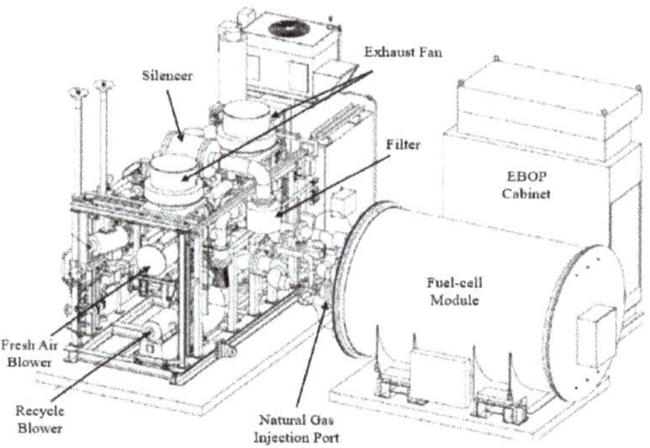

Figure 4. Overall configuration for fuel cell system.

Figure 5 describes the concept of the electricity generation process. The system consists of three modes: A 'heat-up mode' for increasing the initial temperature of the fuel cell stack module, a 'ramp-up mode' for increasing the power output to the rated output, and an 'operating mode' for continuing the rated output.

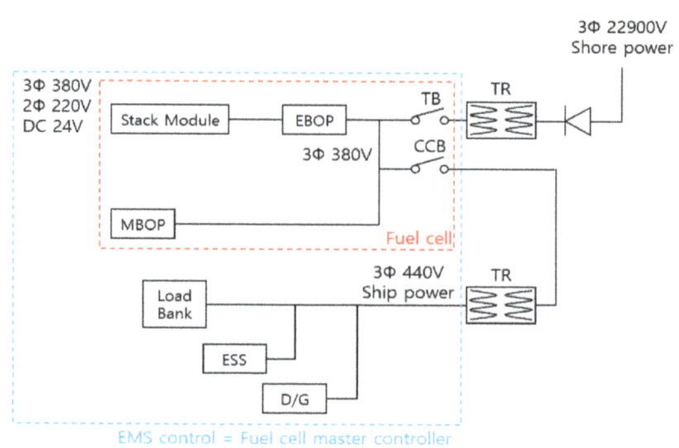

Figure 5. Concept of the fuel cell power generation.

In the 'heat-up' mode of the fuel cell system, as shown in Figure 5, the Terminal Breaker (TB) is closed, and the fuel cell is operated by receiving electricity from the system. Since only the fuel cell consumes power, the Customer Critical Breaker (CCB) is opened. Respectively, it is designed to close the CCB to charge the ESS internal battery. In the 'ramp-up' mode (when the 'heat-up' is completed, and the fuel cell outputs power), the CCB is closed, and the EMS judges whether the ESS is on or off.

3.2.2. Energy Storage System (ESS)

ESS refers to a small/medium-sized electrical storage facility that is to store electrical energy and use it when necessary in aids of a distributed power source in a micro-grid. As shown in Figure 6, ESS comprises the battery and power conditioning system (PCS).

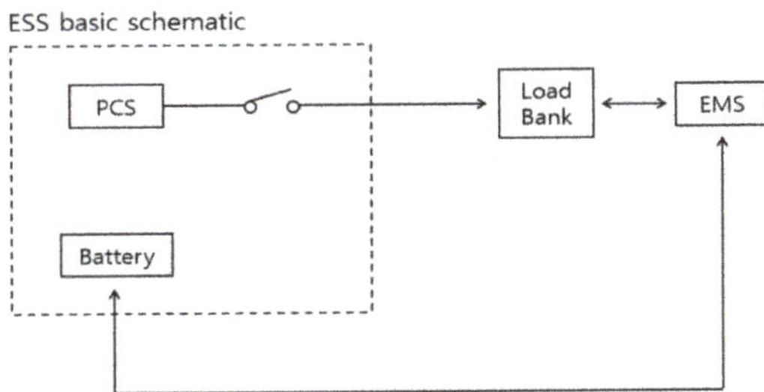

Figure 6. Configuration diagram energy storage system (ESS).

A lead-acid type battery was applied to the test bed. According to Table 2, it was modelled based on the two-way grid connected type.

Table 2. Specification of ESS.

ESS Specification	
Input Voltage	440VAC 3phase 3wire(60 Hz)
Output Voltage	440VAC 3phase 3wire(60 Hz)
Load Capacity	30 kW
Battery Ampere-hour	400 Ah
Inverter Type	Bi-directional grid connected type

The PCS is designed to perform bidirectional power control for DC power and AC power between the grid and the rechargeable battery, to improve the reliability of the power system, and to supply the stored energy quickly at peak power demand.

Therefore, The PCS is to balance active powers from hybrid power systems when the load fluctuation, accidental power supply, or load drop occurs by means of charging or discharging the battery and to contribute to enhancing system stabilization by adjusting the frequency. In addition, it has a functionality to monitor the state of charge (SOC) of the battery in real time and to control the temperature, current and voltage so that it can make the system to be operated with high reliability. It also provides surge protection, automatic overcharge / overload protection as well as overvoltage protection [29,30].

3.2.3. Diesel Generator System

A 50 kW diesel generator used in the test bed is a revolving-field type using a permanent magnet. This system is soundproofed to 75 db or less. As shown in Figure 7, and Tables 3 and 4, it has the synchronous speed of 1800 rpm with four poles.

Figure 7. Appearance of diesel generator.

Table 3. Major specification of diesel generator engine.

Item	Specification
Standby Power Rating	>95 PS
Engine Type	4 stroke, water cooled
Revolution	1800 RPM
Number of Cylinders	6
Fuel	Diesel

Table 4. Major specification of diesel generator.

Item	Specification
Type	Salient Pole Generator
Standby Power Rating	50 kW / 62.5 kVA
Prime Power Rating	45 kW / 56 kVA
Voltage	440 / 254 V
Current	82 A
Phase and Wire	3phase 4wire
Frequency	60 Hz
Number of Pole	4 P
Revolution	1800 RPM
Excitation System	Blushless Self-Exciter

3.2.4. Load Bank

The load bank is a device designed to provide an electrical load for testing power sources, such as generators or uninterruptible power sources. In the case of a load bank used in a test bed, a power line is constructed for load testing, and the current load factor is calculated in the EMS and adjusts the load output from the load bank. The consumption of the voltage and current during this operation was monitored by an analog signal.

As shown in Table 5, the 300 kW load bank adopted the second type of iron chrome (FCHW-2) with high resistivity and low resistance against temperature increase and used a forced air-cooled load bank which was connected in parallel so that the load capacity could be adjusted.

Table 5. Specification of the load bank.

Rated Power	300 Kw				
Rated Voltage	3Ø 440 V, 60 Hz				
Current and resistance value for each capacity					
Unit Capacity (kW)	Quantity	Total Capacity (kW)	Unit Current (A)	Unit Resistance	Cabinet
0.1	1	0.1	0.13	1936Ω*3Ø	
0.2	1	0.2	0.26	968Ω*3Ø	
0.4	1	0.4	0.52	484Ω*3Ø	
0.8	1	0.8	1.04	242Ω*3Ø	
1	1	1	1.31	193.6Ω*3Ø	
2	1	2	2.62	96.8Ω*3Ø	1 box
4	1	4	5.24	48.4Ω*3Ø	
8	1	8	10.49	24.2Ω*3Ø	
16	1	16	20.99	12.1Ω*3Ø	
32	1	32	41.98	6.05Ω*3Ø	
60	4	240	78.78	3.22Ω*3Ø	
Total	15	304.5	399.7	-	

3.2.5. Energy Management System

EMS is a control system configured to monitor the voltage, current, output amount, and system status of each device in real time to operate the system stable. It is to balance the load of the load bank in real time according to the load variations so that each device can be synchronized properly [31–35].

The EMS and each power source—namely MCFC, diesel generator and ESS—are configured to send and receive status and operation commands of the device via the interface as presented in Figure 8.

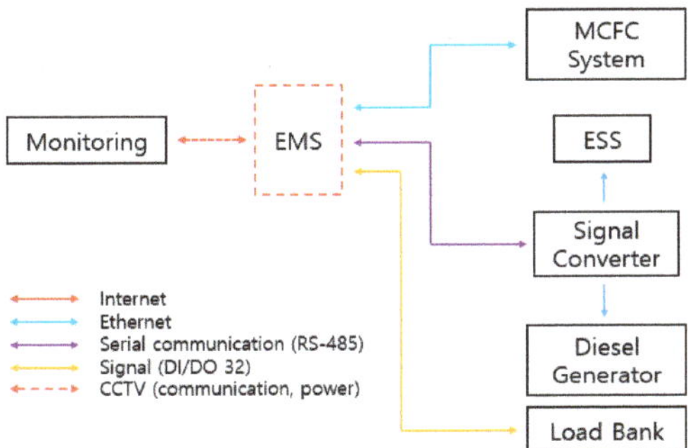

Figure 8. Integrated power control system configuration.

The power source (fuel cell, battery, diesel generator) constituting the hybrid system was controlled by EMS, according to the output range. In the output range of 0–100 kW, the fuel cell produces an output of 100 kW as the base load. When the required load is less than 100 kW, the battery is set to be charged, or the energy is sent to load leveler.

In addition, in the 100–130 kW output period, the battery power is additionally supplied to the load with fuel cell power, and the diesel generator is controlled to operate in the output period of 130 kW or more. When the proposed hybrid system was operated on the basis of the above conditions, the power quality was analyzed by measuring the voltage variation rate and frequency variation rate for each scenario.

4. System Composition for Power Quality Analysis

Figure 9 represents the configuration of MCFC system so that we can monitor and analyses frequency stabilization time as well as the voltage fluctuation rate and frequency fluctuation rate which can occur in the process of synchronizing the other power sources, such as ESS and diesel generator.

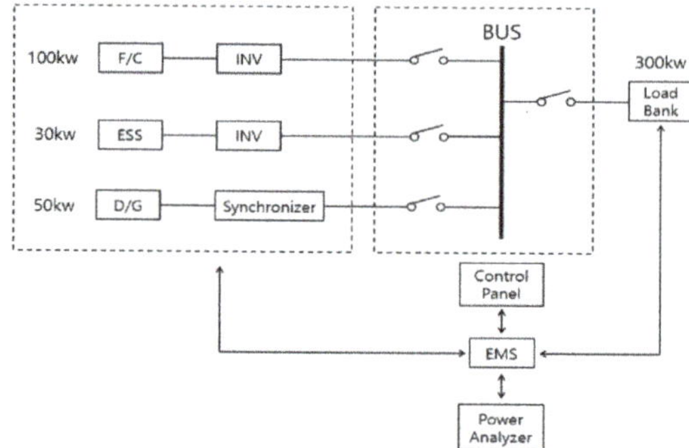

Figure 9. Overall system configuration for power quality analysis.

As shown in the figure, the overall system is designed for power quality analysis which is to measure the voltage, current, power, frequency, and power factor generated during power conversion through the synchronous test, thereby to evaluate the suitability of the combined power source to a marine vessel.

Figure 10 shows the monitoring system configuration for power quality analysis. We could monitor power data (voltage, current, frequency, voltage variation, frequency variation, apparent/active/reactive power and power factor) in real time and store them on the LabVIEW interface.

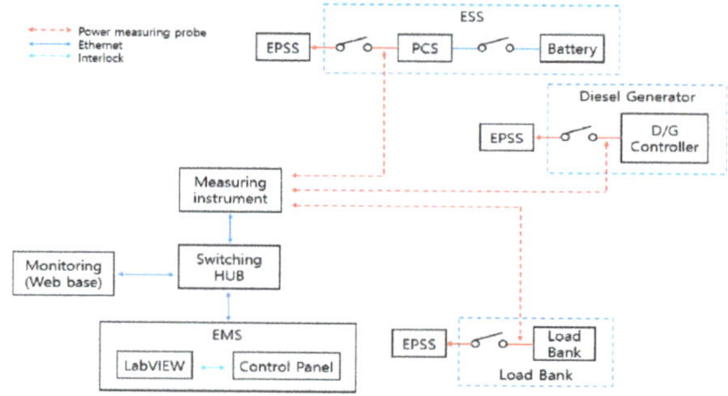

Figure 10. Monitoring system configuration for power quality analysis.

Twelve channels of the measuring points were assigned on MCFC, ESS and secondary power supply R, T of diesel generator. A high-voltage probe (TP0200) was used for the measurement as pictured in Figure 11, whereas a current measuring clamp i410 type was applied, as shown in Figure 12.

Figure 11. Installed voltage measuring equipment.

Figure 12. Installed current measuring equipment.

5. Development of Load Scenario for Power Quality Analysis of Fuel Cell-Based Marine Hybrid Power System

5.1. Load Scenario Development

In this study, when the MCFC, which is responsible for the base load of the fuel cell-based combined power source, is applied to the ship with other power sources, such as ESS and diesel generator, we developed a load scenario for power quality analysis through voltage fluctuation rate, frequency fluctuation rate, and frequency stabilization time for power faults that may occur in the process of synchronizing to or disconnection from a single bus line. Scenario requested sailing information of the 5,500 TEU vessel of 'H' shipping company and collected and analyzed the data.

Based on the data of main engine and generator output of 5,500 TEU class container ship, we analyzed the load variation characteristics in normal seagoing operation in order to investigate whether each power source can maintain proper power quality for load fluctuation, such as synchronization and system deviation. We have developed a Load Scenario to determine if it is possible.

Using the real-time operation data, the load case was developed based on the following steps.

- Analyzes the load pattern according to the size of the fluctuating load when analyzing operational data.

- Exclude generator load which is 2.5–5% of the main engine loads as documented in IMO MEPC.1 / Circ.681
- Develop Load Scenario for verifying load followability of the fuel cell-based ship's hybrid power systems according to load variation
- For cargo ships with the main engine power of 10000 kW or above, P_{AE} is defined as Equation (1);

$$\mathrm{P}_{AE(MCRME>10000kW)} = (0.025\ X \sum_{i=1}^{nME} MCR_{MEi}) + 250. \tag{1}$$

- For cargo ships with the main engine power below 10000 kW, P_{AE} is defined as equation (2);

$$\mathrm{P}_{AE(MCRME<10000kW)} = (0.05\ X \sum_{i=1}^{nME} MCR_{MEi}). \tag{2}$$

5.2. Development of Load Scenario Using Real-Time Operation Load Pattern

5.2.1. Sync Phase and Conditions for Normal Seagoing Case 1

Figure 13 and Table 6 summarize the synchronization and off-gird of the load pattern of the normal seagoing case 1 with the peak load of 127 kW of the combined power source. ESS is configured to discharge after one synchronization at a time when the total load of the complex power source is over 100 kW, and then it is changed to the charging mode by the ESS SOC reference value, and the charging is progressed in the state connected to the system.

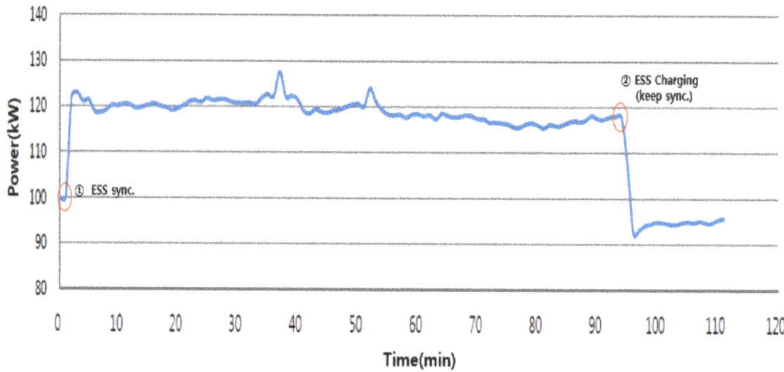

Figure 13. Synchronization and system deviation for load pattern of normal seagoing case 1.

Table 6. Synchronization condition in normal seagoing case 1.

Status	Power Sources
① ESS sync.	MCFC + ESS
② ESS Charging (keep sync.)	MCFC + ESS

5.2.2. Sync Phase and Conditions for Normal Seagoing Case 2

Figure 14 and Table 7 summarize the synchronization and out of gird areas when the load of the combined power source reaches 140 kW according to the load pattern of the normal seagoing case 2. The ESS is synchronized with the MCFC at the moment when the total load exceeds 100 kW. The diesel generator is synchronized with the system at a total load of 130 kW or more, and is configured so that when the load drops below 130 kW, the system disengages. When the load drops below 100 Kw, the

ESS is also separated from the MCFC, and the scenario is configured to synchronize with the ESS as soon as the total load rises above 100 kW again.

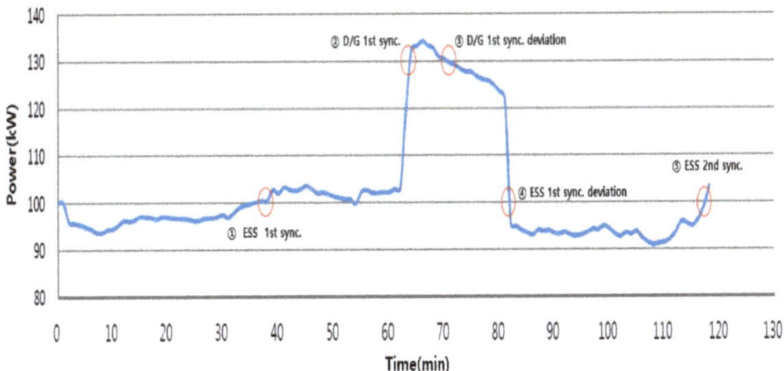

Figure 14. Synchronization and system deviation for load pattern of normal seagoing case 2.

Table 7. Synchronization condition in normal seagoing case 2.

Status	Power Sources
① ESS 1st sync.	MCFC + ESS
② D/G 1st sync.	MCFC + ESS+ D/G
③ D/G 1st sync. deviation	MCFC + ESS
④ ESS 1st sync. deviation	MCFC
⑤ ESS 2nd sync.	MCFC + ESS

5.2.3. Sync Phase and Conditions for Normal Seagoing Case 3

Figure 15 and Table 8 summarize the synchronization and off grid part of the normal seagoing case 3. In the ESS, there are two synchronization periods and system outages repeatedly according to the load of the combined power source. In the section where the charge interval and the discharge interval are repeated, charging and discharging are performed in a state, in which the system synchronization is maintained. The diesel generator has two synchronization periods and a system outage period.

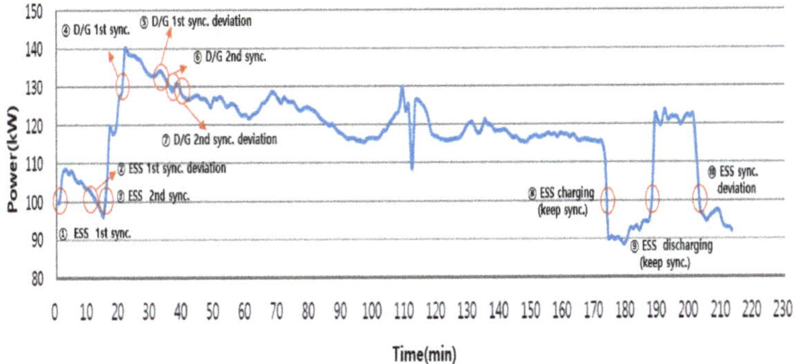

Figure 15. Synchronization and system deviation for load pattern of normal seagoing case 3.

Table 8. Synchronization condition in normal seagoing case 3.

Status	Power Sources
① ESS 1st sync.	MCFC + ESS
③ ESS 1st sync. deviation	MCFC
④ ESS 2nd sync.	MCFC + ESS
⑤ D/G 1st sync.	MCFC + ESS + D/G
⑥ D/G 1st sync. deviation	MCFC +ESS
⑦ D/G 2nd sync.	MCFC + ESS + D/G
⑧ D/G 2nd sync. Deviation	MCFC + ESS
⑨ ESS Charging (keep sync.)	MCFC + ESS
⑩ ESS Discharging (keep sync.)	MCFC + ESS
⑪ ESS sync. Deviation	MCFC

6. Procedures for Power Quality Test for Ship Hybrid Power System

6.1. Synchronization Procedure between Hybrid Power Sources

6.1.1. The procedure for ESS Synchronization

ESS system voltage is followed by priority, and when the system voltage is transmitted to the power conversion device (PCS) of the ESS via the sensor, the voltage is controlled by following the system voltage.

When the synchronous signal is transmitted from the EMS to the PCS, the PCS follows the system frequency. The PCS adjusts the frequency so that the system can be stably maintained. In this case, the PCS controls the system using the current and perform charging and discharging actions in accordance with the pre-setting power outputs.

The ESS synchronization time would be approximately 0.5 seconds during which PCS turns on, and EMS generates a synchronization signal in the normal operation state in which charging and discharging are enabled.

6.1.2. Procedure of Diesel Generator

When the synchronization signal is produced to the EMS in normal operation, the diesel generator adjusts the voltage based on the grid voltage and synchronizes with the system frequency. The synchronization controller of the diesel generator plays a role in maintaining the system stable by adjusting the frequency. At this time, the amount of electric power in the form of current control mode is limited, so that more current of electric power than the pre-set can be prevented from flowing to the system.

Once the EMS synchronization signal is generated in the normal operation state of the diesel generator, it takes about 3–4 seconds to synchronize to the system.

6.2. Power Measurement Point and Selection Criteria When Synchronizing Hybrid Power Sources

The effect of voltage and frequency on the system was investigated by measuring the synchronization time between the ship's combined power source in two times: Two seconds before synchronization and five seconds after synchronization. The measuring time was established based on Korean Classification Rule. Therefore, the built-in LabVIEW-based power analysis system for test bed experiment can store one power measurement data every 33.3 ms.

The ESS was configured to measure two seconds before synchronization and five seconds after synchronization in consideration of the short synchronization time; it could be confirmed through data that ESS took 7–10 seconds depending on the system condition when synchronizing.

The synchronization time of the diesel generator was designed to measure two seconds before synchronization and five seconds after synchronization in the same manner as ESS. However, in case of diesel generator, it took about 10–12 seconds because the synchronization time usually took 3–5 seconds, which was confirmed through the results.

7. Analysis Result of Power Quality Data

7.1. Target of Voltage, Frequency Variation and Stabilization Time

With regard to synchronization of MCFC, ESS and diesel generation, the voltage, frequency variation, and frequency variation and stabilization time in the transient state of disconnection from the grid were tested with reference to the Korean Register of Shipping, "Class 6 Electrical Equipment and Control Systems" in Table 9 [36].

Table 9. Rules for the classification of electrical installation and control systems of the Korean Register of Shipping.

Item	Regulation		Stabilization Time
	Steady State	Transient State	
Frequency	±5%	±10% (5s)	<15s
Voltage	+6%, −10%	±20% (1.5s)	-

7.2. Test Results of Synchronization Interval and Breakaway Interval

7.2.1. Test Results for ESS Synchronization and Breakaway

Figure 16 shows the voltage fluctuation trend of the ESS before and after the synchronization in the MCFC system. It can be found that the system voltage before the ESS synchronization encountered a slight deviation, but the follow-up control was effectively confirmed after the ESS synchronization.

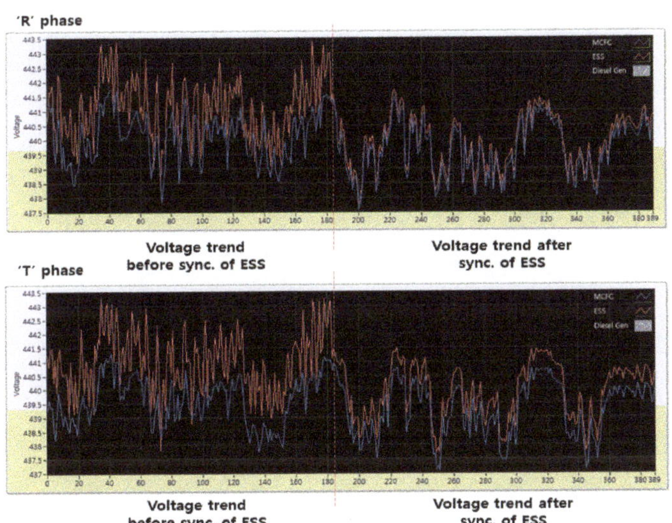

Figure 16. Voltage change during ESS synchronization.

In Figure 17, the synchronous signal from IMES followed the system voltage and frequency. The reference frequency for both MCFC and ESS was 60 HZ, and that was exactly matched each other after synchronization.

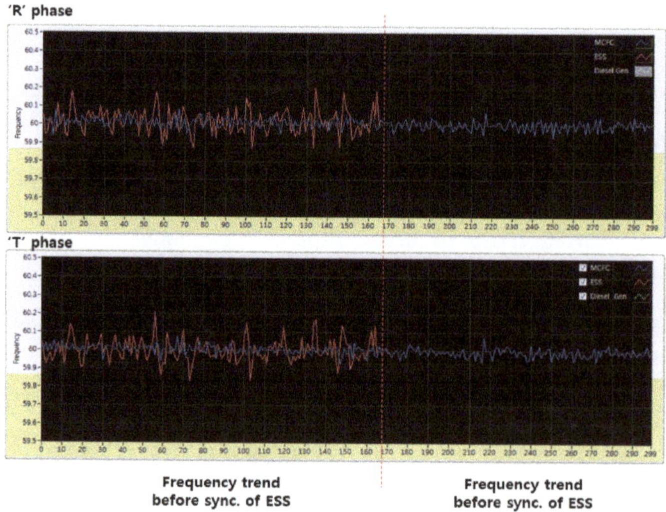

Figure 17. Frequency change during ESS synchronization.

When a system breakaway signal was generated from the IMES, the ESS was disconnected from the system. As shown in Figure 18, the separation time of the ESS from the system was estimated less than 0.5 seconds, which was the same as the synchronization time. The voltage fluctuation trend of the ESS, before and after synchronization in the MCFC system, was also clearly presented in the figure.

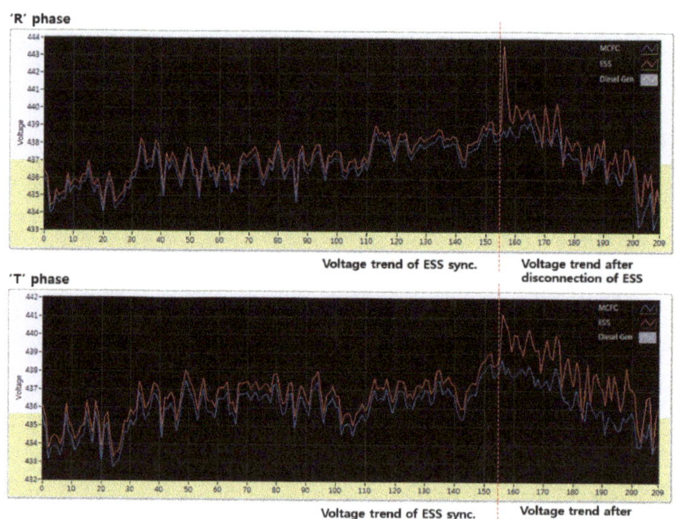

Figure 18. Voltage change when the ESS system was disconnected.

Figure 19 shows that the parallel operation of the MCFC and the ESS and the breakaway proceedings when the system breakaway signal was sent from the IMES to the ESS. In this case, the frequency was observed to be slightly hunt, but it did not affect the existing system.

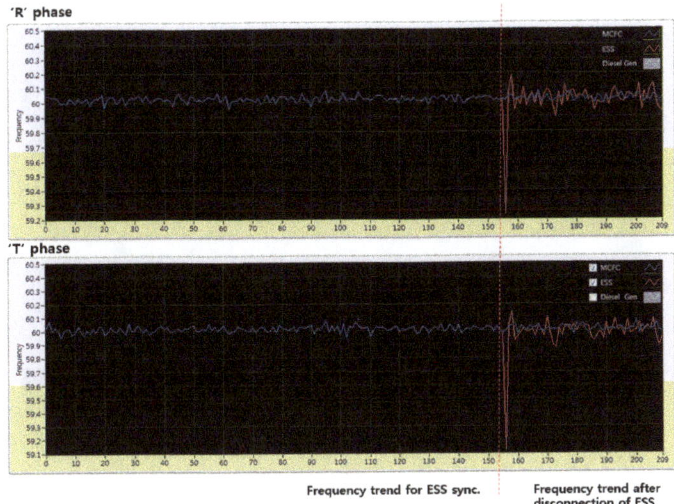

Figure 19. Frequency change in ESS system deviation.

7.2.2. Test Results For Diesel Generator Synchronization and Breakaway

Figure 20 shows the voltage fluctuation trend before and after the synchronization of the diesel generator with the MCFC and ESS systems. When the EMS sent the synchronization signal, it followed the system voltage so that the voltage could be synchronized—thereby, the system voltage was exactly matched.

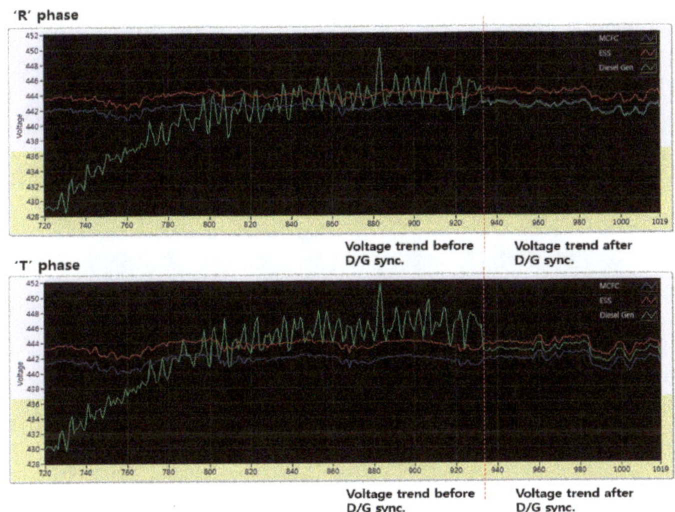

Figure 20. Voltage change when synchronizing diesel generators.

Figure 21 shows the frequency fluctuation trend before and after synchronizing the diesel generator with the MCFC and ESS systems. It revealed that once the EMS sent the diesel generator synchronization signal, it started to follow the system voltage and achieved the voltage synchronization. Thereafter, it followed the system frequency and attained frequency synchronization.

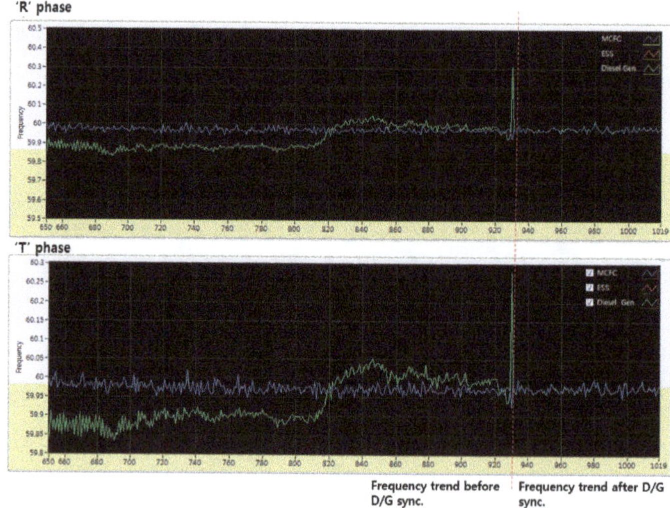

Figure 21. Frequency change when synchronizing diesel generators.

When the system separation signal was generated from the EMS, the diesel generator was disconnected from the system by the synchronization switch fitted in the synchronization controller. In Figure 22, a waveform similar to the synchronous voltage waveform was plotted in the same manner as in the system state before synchronizing the diesel generator.

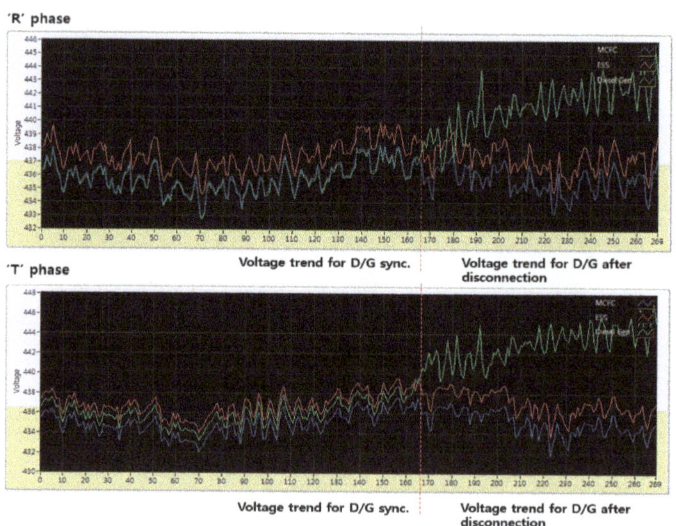

Figure 22. Voltage change when diesel generators are broken away.

Figure 23 shows the synchronous operation of all power systems in parallel. When the system breakaway signal went off to the IG-NT from the IMES, the frequency was observed to be a hunt, which occurred after the breakaway, thereby such an outage had no effect on the existing system.

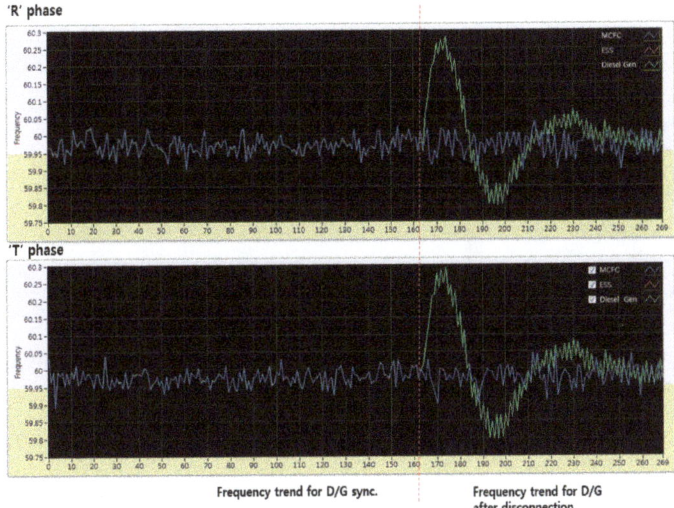

Figure 23. Frequency change when leaving diesel generator system.

7.3. *Analysis of Frequency Variation Rate and Frequency Stabilization Time Data Based on Load Scenarios*

7.3.1. Power Data before and after Synchronization for Normal Seagoing Case 1

Table 10 shows the analysis results pertinent to the voltage and frequency fluctuation measured in the load pattern of the normal seagoing case 1.

Table 10. Power data analysis before and after ESS synchronization.

Item	Voltage Fluctuation						Frequency Fluctuation					
	MCFC		ESS		D/G		MCFC		ESS		D/G	
Condition of Synchronization	Bef	Aft	Bef	Aft	Bef	Aft	Bef	Aft	Bef	Aft	Bef	Aft
Maximum	440.8	440.8	443.7	444.0	-	-	60.1	60.1	60.2	60.2	-	-
Minimum	438.4	438.3	440.1	438.5	-	-	60.0	59.9	59.8	59.9	-	-
Average	439.6	439.7	441.8	440.2	-	-	60.0	60.0	60.0	60.0	-	-
Maximum rate of change (%)	1.00	1.00	1.01	1.01	-	-	1.00	1.00	1.00	1.00	-	-
Minimum rate of change (%)	1.00	1.00	1.00	1.00	-	-	1.00	1.00	1.00	1.00	-	-
Regulation (%)	±1.00	±1.00	±1.00	±1.00	-	-	±1.00	±1.00	±1.00	±1.00	-	-

When the ESS was synchronized with the system, the voltage fluctuation rate was within ±1.01%, and the voltage fluctuation rate ranged from +6 to −10 %. Therefore, it could be confirmed that the voltage fluctuation hardly occurred during system synchronization, and the voltage characteristic was kept stable.

Likewise, the stability of frequency was also confirmed since the frequency variation rate was within ±1.01%, much lower than the standard of ±5%.

7.3.2. Power Data before and after Synchronization for Normal Seagoing Case 2

Tables 11–14 describe the analysis results from power data measured during synchronization and breakaway of the diesel generator under the normal seagoing case 2.

Table 11. Power data analysis before and after ESS synchronization.

Item	Voltage Fluctuation						Frequency Fluctuation					
	MCFC		ESS		D/G		MCFC		ESS		D/G	
Condition of Synchronization	Bef	Aft	Bef	Aft	Bef	Aft	Bef	Aft	Bef	Aft	Bef	Aft
Maximum	439.6	439.5	445.0	440.7	-	-	60.1	60.1	60.1	60.4	-	-
Minimum	437.2	437.5	440.8	437.8	-	-	60.0	60.0	0.0	60.0	-	-
Average	438.6	438.5	442.5	438.8	-	-	60.0	60.0	8.1	60.0	-	-
Maximum rate of change (%)	1.00	1.00	1.01	1.00	-	-	1.00	1.00	1.00	1.01	-	-
Minimum rate of change (%)	0.99	1.09	1.00	1.09	-	-	1.00	1.00	0.00	1.00	-	-
Regulation (%)	±1.00	±1.00	±1.01	±1.00	-	-	±1.00	±1.00	±0.13	±1.00	-	-

Table 12. Power data analysis before and after diesel generator synchronization.

Item	Voltage Fluctuation						Frequency Fluctuation					
	MCFC		ESS		D/G		MCFC		ESS		D/G	
Condition of Synchronization	Bef	Aft	Bef	Aft	Bef	Aft	Bef	Aft	Bef	Aft	Bef	Aft
Maximum	441.1	440.7	442.8	442.4	445.1	444.6	60.1	60.1	60.1	60.1	60.0	60.2
Minimum	438.0	437.9	439.6	439.6	426.9	438.2	60.0	60.0	60.0	60.0	59.8	59.9
Average	440.0	439.6	441.6	441.3	434.8	440.5	60.0	60.0	60.0	60.0	59.9	60.0
Maximum rate of change (%)	1.00	1.00	1.01	1.01	1.01	1.01	1.00	1.00	1.00	1.00	1.00	1.00
Minimum rate of change (%)	1.00	1.00	1.00	1.00	0.97	1.00	1.00	1.00	1.00	1.00	1.00	1.00
Regulation (%)	±1.00	±1.00	±1.00	±1.00	±0.99	±1.00	±1.00	±1.00	±1.00	±1.00	±1.00	±1.00

Table 13. Power data analysis before and after departing from the diesel generator system.

Item	Voltage Fluctuation						Frequency Fluctuation					
	MCFC		ESS		D/G		MCFC		ESS		D/G	
Condition of Synchronization	Bef	Aft	Bef	Aft	Bef	Aft	Bef	Aft	Bef	Aft	Bef	Aft
Maximum	438.5	439.2	440.2	440.9	438.5	444.4	60.1	60.1	60.1	60.1	60.1	60.4
Minimum	434.5	436.2	436.2	437.9	434.5	436.0	60.0	59.9	60.0	59.9	60.0	59.9
Average	437.2	437.6	438.9	439.3	437.2	439.6	60.0	60.0	60.0	60.0	60.0	60.0
Maximum rate of change (%)	1.00	1.00	1.00	1.00	1.00	1.01	1.00	1.00	1.00	1.00	1.00	1.01
Minimum rate of change (%)	0.99	0.99	0.99	1.00	0.99	0.99	1.00	1.00	1.00	1.00	1.00	1.00
Regulation (%)	±0.99	±0.99	±1.00	±1.00	±0.99	±1.00	±1.00	±1.00	±1.00	±1.00	±1.00	±1.00

Table 14. Power data analysis before and after ESS system outage.

Item	Voltage Fluctuation						Frequency Fluctuation					
	MCFC		ESS		D/G		MCFC		ESS		D/G	
Condition of Synchronization	Bef	Aft	Bef	Aft	Bef	Aft	Bef	Aft	Bef	Aft	Bef	Aft
Maximum	441.2	440.7	443.3	441.5	-	-	60.1	60.1	60.3	60.1	-	-
Minimum	438.1	437.2	438.4	438.8	-	-	60.0	60.0	60.0	59.9	-	-
Average	439.9	439.2	440.3	440.4	-	-	60.0	60.0	60.0	60.0	-	-
Maximum rate of change (%)	1.00	1.00	1.01	1.00	-	-	1.00	1.00	1.00	1.00	-	-
Minimum rate of change (%)	1.00	0.99	1.00	1.00	-	-	1.00	1.00	1.00	1.00	-	-
Regulation (%)	±1.00	±1.00	±1.00	±1.00	-	-	±1.00	±1.00	±1.00	±1.00	-	-

The voltage fluctuation rate was observed within ±1.01% when the ESS and the diesel generator were synchronized to the system. Since the voltage fluctuation was placed in the standard range of +6 to −10 %, the system stability was verified for this case as well.

The frequency deviation was within ±1.01%. In the same way, given the standard range of ±5%, frequency stability has been confirmed.

7.3.3. Power Data before and after Synchronization for Normal Seagoing Case 3

The analysis results for the voltage and frequency variation measured in the load pattern of normal seagoing case 3 are shown across Tables 15–18 which deal with power data for ESS synchronization and breakaway, as well as that for the diesel generator.

Table 15. Power data analysis before and after ESS synchronization.

Item	Voltage Fluctuation						Frequency Fluctuation					
	MCFC		ESS		D/G		MCFC		ESS		D/G	
Condition of Synchronization	Bef	Aft	Bef	Aft	Bef	Aft	Bef	Aft	Bef	Aft	Bef	Aft
Maximum	442.9	442.6	445.2	442.9	-	-	60.1	60.1	60.2	60.1	-	-
Minimum	440.4	440.3	440.6	440.5	-	-	60.0	60.0	59.9	60.0	-	-
Average	441.7	441.5	442.7	441.8	-	-	60.0	60.0	60.0	60.0	-	-
Maximum rate of change (%)	1.01	1.01	1.01	1.01	-	-	1.00	1.00	1.00	1.00	-	-
Minimum rate of change (%)	1.00	1.00	1.00	1.00	-	-	1.00	1.00	1.00	1.00	-	-
Regulation (%)	±1.00	±1.00	±1.01	±1.00	-	-	±1.00	±1.00	±1.00	±1.00	-	-

Table 16. Power data analysis before and after ESS system outage.

Item	Voltage Fluctuation						Frequency Fluctuation					
	MCFC		ESS		D/G		MCFC		ESS		D/G	
Condition of Synchronization	Bef	Aft	Bef	Aft	Bef	Aft	Bef	Aft	Bef	Aft	Bef	Aft
Maximum	441.3	441.4	441.6	443.5	-	-	60.0	60.0	60.0	61.3	-	-
Minimum	439.0	438.7	439.4	438.2	-	-	60.0	60.0	60.0	59.8	-	-
Average	440.4	440.3	440.8	440.8	-	-	60.0	60.0	60.0	60.0	-	-
Maximum rate of change (%)	1.00	1.00	1.00	1.01	-	-	1.00	1.00	1.00	1.02	-	-
-Minimum rate of change (%)	1.00	1.00	1.00	1.00	-	-	1.00	1.00	1.00	1.00	-	-
Regulation (%)	±1.00	±1.00	±1.00	±1.00	-	-	±1.00	±1.00	±1.00	±1.00	-	-

Table 17. Power data analysis before and after diesel generator synchronization.

Item	Voltage Fluctuation						Frequency Fluctuation					
	MCFC		ESS		D/G		MCFC		ESS		D/G	
Condition of Synchronization	Bef	Aft	Bef	Aft	Bef	Aft	Bef	Aft	Bef	Aft	Bef	Aft
Maximum	437.3	437.7	438.9	439.3	444.6	443.5	60.1	60.1	60.1	60.1	60.2	60.5
Minimum	432.8	433.2	434.4	434.8	426.4	433.0	59.9	59.9	59.9	59.9	59.7	59.9
Average	435.5	435.7	437.1	437.3	434.8	436.7	60.0	60.0	60.0	60.0	59.9	60.0
Maximum rate of change (%)	0.99	0.99	1.00	1.00	1.01	1.01	1.00	1.00	1.00	1.00	1.00	1.01
Minimum rate of change (%)	0.98	0.98	0.99	0.99	0.97	0.98	1.00	1.00	1.00	1.00	1.00	1.00
Regulation (%)	±0.99	±0.99	±0.99	±0.99	±0.99	±0.99	±1.00	±1.00	±1.00	±1.00	±1.00	±1.00

Table 18. Power data analysis before and after diesel generators outage.

Item	Voltage Fluctuation						Frequency Fluctuation					
	MCFC		ESS		D/G		MCFC		ESS		D/G	
Condition of Synchronization	Bef	Aft	Bef	Aft	Bef	Aft	Bef	Aft	Bef	Aft	Bef	Aft
Maximum	439.1	439.7	440.7	441.2	439.0	444.8	60.1	60.1	60.1	60.1	60.1	60.3
Minimum	435.3	436.1	436.9	437.7	435.2	436.1	60.0	59.9	60.0	59.9	60.0	59.9
Average	437.4	437.7	439.0	439.3	437.3	439.3	60.0	60.0	60.0	60.0	60.0	60.0
Maximum rate of change (%)	1.00	1.00	1.00	1.00	1.00	1.01	1.00	1.00	1.00	1.00	1.00	1.01
Minimum rate of change (%)	0.99	0.99	0.99	0.99	0.99	0.99	1.00	1.00	1.00	1.00	1.00	1.00
Regulation (%)	±0.99	±0.99	±1.00	±1.00	±0.99	±1.00	±1.00	±1.00	±1.00	±1.00	±1.00	±1.00

Since when synchronized to ESS and diesel generators, voltage stability was confirmed with the voltage fluctuation rate measured between ± 1.01 % (standard range of +6 to −10%).The frequency variation was measured within ± 1.01 %—therefore, it can be seen that frequency fluctuation hardly occurs during system synchronization.

During the breakaways of ESS and diesel generator, it was confirmed that the voltage and frequency were not subjected to the deviation from the acceptable ranges.

8. Conclusions

Until now, most of the marine fuel cell research works have been limited to concentrating on the low power capacity of PEMFCs for small vessels or MCFCs for medium and large-sized vessels as auxiliary power sources.

However, in this paper, a hybrid power system was constructed along with a battery and a generator system, in order to apply the MCFC as the main power source for medium and large-sized ships. Their performance in parallel operation was monitored and investigated based on the actual experiment with the test bed.

Experiment results revealed that the deviation levels of voltage and frequency were kept within the standard ranges across all operational cases. Therefore, system synchronization was proven stable. It also confirmed that power quality of the built-in hybrid power source the test bed was compliant with the rules of the Korean Register of Shipping.

Therefore, the novelty of this research can be placed on this contribution to the design and the application of the hybrid power system using MCFC for the propulsion power system of the middle and large-sized ships.

Author Contributions: Conceptualization, K.Y. and H.J.; Methodology, K.Y., H.J. and S.K.; Formal analysis, H.J. and S.K.; Resources, H.J. and S.K.; X.X.; Writing—original draft preparation, K.Y. and H.J.; Writing—review and editing, K.Y., H.J. and S.K.; Visualization, H.J. and S.K.; Supervision, K.Y.

Funding: This research received no external funding.

Acknowledgments: We thank our colleagues from Kido Park, who provided insight and expertise that greatly assisted the research, although they may not agree with all of the conclusions of this paper. The authors would also like to thank POSCO Engineering co. for approval to use 'DFC300' model.

Conflicts of Interest: The authors declare no conflict of interest.

References

1. DNVGL. Maritime Fuel Cell Applications—Technologies and Ongoing Developments. Available online: http://www.jterc.or.jp/koku/koku_semina/pdf/180221_presentation-01.pdf (accessed on 21 February 2018).
2. Fuel Cells Bulletin. ABB, SINTEF Testing Viability of Fuel Cells for Ship Propulsion. Available online: https://reader.elsevier.com/reader/sd/pii/S1464285918304504?token=75B63CA01658C4DDB95A2A3F83B391DF39A34D3FE9AC3C4D2FED982491E063E445CD4A2719E654939A296BD3B8BBE9DB (accessed on 1 February 2017).

3. Andújar, J.M.; Segura, F. Fuel Cells: History and Updating. A Walk along Two Centuries. *Renew. Sustain. Energy Rev.* **2009**, *13*, 2309–2322. [CrossRef]
4. Pestana, H. Future Trends of Electrical Propulsion and Implications to Ship Design. *Marit. Technol. Eng.* **2014**, *478*, 797–803.
5. Larminie, J.; Dicks, A. *Fuel Cell Systems Explained*; Wiley: Hoboken, NJ, USA, 2003.
6. McConnell, V.P. Now, Voyager? The Increasing Marine Use of Fuel Cells. *Fuel Cells Bull.* **2010**, *1*, 12–17. [CrossRef]
7. Yuan, J.; Ng, S.H.; Sou, W.S. Uncertainty Quantification of CO 2 Emission Reduction for Maritime Shipping. *Energy Policy* **2016**, *88*, 113–130. [CrossRef]
8. Biert, L.V.; Godjevac, M.; Visser, K.; Aravind, P.V. A Review of Fuel Cell Systems for Maritime 490 Applications. *J. Power Sources* **2016**, *327*, 345–364. [CrossRef]
9. De-Troya, J.J.; Álvarez, C.; Fernández-Garrido, C.; Carral, L. Analysing the Possibilities of Using Fuel 492 Cells in Ships. *Int. J. Hydrog. Energy* **2016**, *41*, 2853–2866. [CrossRef]
10. Wee, J.H. Contribution of Fuel Cell Systems to CO2 Emission Reduction in Their Application Fields. *Renew. Sustain. Energy Rev.* **2010**, *14*, 735–744. [CrossRef]
11. Tronstad, T.; Åstrand, H.H.; Haugom, G.P.; Langfeldt, L. Study on the Use of Fuel Cells in Shipping. *DNVGL* **2017**, *1*, 1–108.
12. Vogler, F.; Würsig, G. New Developments for Maritime Fuel Cell Systems. In Proceedings of the World Hydrogen Energy Conference, Essen, Germany, 16–21 May 2010.
13. Welaya, Y.M.A.; El Gohary, M.M.; Ammar, N.R. A Comparison between Fuel Cells and Other Alternatives for Marine Electric Power Generation. *Int. J. Nav. Archit. Ocean Eng.* **2011**, *3*, 141–149. [CrossRef]
14. Mekhilef, S.; Saidur, R.; Safari, A. Comparative Study of Different Fuel Cell Technologies. *Renew. Sustain. Energy Rev.* **2012**, *16*, 981–989. [CrossRef]
15. Schneider, J.; Dirk, S.; Motor, P. ZEMShip. In Proceedings of the World Hydrogen Energy Conference, Essen, Germany, 16–21 May 2010.
16. Mehta, V.; Cooper, J.S. Review and Analysis of PEM Fuel Cell Design and Manufacturing. *J. Power Sources* **2003**, *114*, 32–53. [CrossRef]
17. Hamilton, P.J.; Pollet, B.G. Polymer Electrolyte Membrane Fuel Cell (PEMFC) Flow Field Plate: Design, Materials and Characterisation. *Fuel Cells* **2010**, *10*, 489–509. [CrossRef]
18. Erdinc, O.; Uzunoglu, M. Recent Trends in PEM Fuel Cell-Powered Hybrid Systems: Investigation of Application Areas, Design Architectures and Energy Management Approaches. *Renew. Sustain. Energy Rev.* **2010**, *14*, 2874–2884. [CrossRef]
19. Innogy. Green Methanol as CO2-Neutral, Renewable Energy Carrier. Available online: https://www.engerati.com/system/files/presentationpdf/175.pdf (accessed on 4 October 2017).
20. Huang, X.; Zhang, Z.; Jiang, J. Fuel Cell Technology for Distributed Generation: An Overview F. C. Distributed Generation: An Overview. In Proceedings of the International Symposium on Industrial Electronicsm, Montreal, QC, Canada, 9–12 July 2006.
21. Inal, O.; Deniz, C. Fuel Cell Availability for Merchant Ships Fuel Cell Availability for Merchant Ships. In Proceedings of the International Symposium on Naval Architecture and Maritime, Istanbul, Turkey, 24–25 April 2018.
22. Martinić, F.; Radica, G.; Barbir, F. Application and Analysis of Solid Oxide Fuel Cells in Ship Energy Systems. *Brodogradnja* **2018**, *69*, 53–68. [CrossRef]
23. Varga, A. *Introduction to Fuel Cell Technology*; Springer: Boston, MA, USA, 2007; pp. 1–32.
24. Inci, M.; Türksoy, Ö. Review of Fuel Cells to Grid Interface: Configurations, Technical Challenges and Trends. *J. Clean. Prod.* **2019**, *213*, 1353–1370. [CrossRef]
25. Milewski, J.; Świercz, T.; Badyda, K.; Miller, A.; Dmowski, A.; Biczel, P. The Control Strategy for a Molten Carbonate Fuel Cell Hybrid System. *Int. J. Hydrog. Energy* **2010**, *35*, 2997–3000. [CrossRef]
26. Neburchilov, V.; Martin, J.; Wang, H.; Zhang, J. A Review of Polymer Electrolyte Membranes for Direct Methanol Fuel Cells. *J. Power Sources* **2007**, *169*, 221–238. [CrossRef]
27. Tomczyk, P. MCFC versus Other Fuel Cells-Characteristics, Technologies and Prospects. *J. Power Sources* **2006**, *160*, 858–862. [CrossRef]
28. FuelCell Energy Inc. 300kW DFC300 Specification. Available online: http://www.fuelcellenergy.com/assets/PID000155_FCE_DFC300_r1_hires.pdf (accessed on 16 August 2011).

29. Han, J.; Charpentier, J.F.; Tang, T. An Energy Management System of a Fuel Cell/Battery Hybrid Boat. *Energies* **2014**, *7*, 2799–2820. [CrossRef]
30. Choi, C.H.; Yu, S.; Han, I.S.; Kho, B.K.; Kang, D.G.; Lee, H.Y.; Seo, M.S.; Kong, J.W.; Kim, G.; Ahn, J.W.; et al. Development and Demonstration of PEM Fuel-Cell-Battery Hybrid System for Propulsion of Tourist Boat. *Int. J. Hydrog. Energy.* **2016**, *41*, 3591–3599. [CrossRef]
31. San Martín, J.I.; Zamora, I.; Asensio, F.J.; García Villalobos, J.; San Martín, J.J.; Aperribay, V. Control and Management of a Fuel Cell Microgrid. Energy Efficiency Optimization. *Renew. Energy Power Qual. J.* **2017**, *7*, 314–319. [CrossRef]
32. Kirubakaran, A.; Jain, S.; Nema, R.K. A Review on Fuel Cell Technologies and Power Electronic Interface. *Renew. Sustain. Energy Rev.* **2009**, *13*, 2430–2440. [CrossRef]
33. Lotfy, M.E.; Senjyu, T.; Farahat, M.A.F.; Abdel-Gawad, A.F.; Matayoshi, H. A Polar Fuzzy Control Scheme for Hybrid Power System Using Vehicle-to-Grid Technique. *Energies* **2017**, *10*, 1083. [CrossRef]
34. Bassam, A.M.; Phillips, A.B.; Turnock, S.R.; Wilson, P.A. An Improved Energy Management Strategy for a Hybrid Fuel Cell/Battery Passenger Vessel. *Int. J. Hydrog. Energy* **2016**, *41*, 22453–22464. [CrossRef]
35. Bassam, A.M.; Phillips, A.B.; Turnock, S.R.; Wilson, P.A. Development of a Multi-Scheme Energy Management Strategy for a Hybrid Fuel Cell Driven Passenger Ship. *Int. J. Hydrog. Energy* **2017**, *42*, 623–635. [CrossRef]
36. Park, S.-K.; Youn, Y.-M. Analysis of International Standardization Trend for the Application of Fuel Cell Systems on Ships. *J. Korean Soc. Mar. Environ. Saf.* **2015**, *20*, 579–585. [CrossRef]

Journal of
Marine Science and Engineering

Article

Comparison and Verification of Reliability Assessment Techniques for Fuel Cell-Based Hybrid Power System for Ships

Hyeonmin Jeon [1], Kido Park [1,2] and Jongsu Kim [1,*]

1 Department of Marine System Engineering, Korea Maritime and Ocean University, Busan 49112, Korea; jhm861104@kmou.ac.kr (H.J.); kdpark@krs.co.kr (K.P.)
2 Future Technology Research Team, Korea Resister, Busan 46762, Korea
* Correspondence: jongskim@kmou.ac.kr; Tel.: +82-(0)102-558-1197

Received: 11 December 2019; Accepted: 22 January 2020; Published: 24 January 2020

Abstract: In order to secure the safe operation of the ship, it is crucial to closely examine the suitability from the design stage of the ship, and to set up a preliminary review and countermeasures for failures and defects that may occur during the construction process. In shipyards, the failure mode and effects analysis (FMEA) evaluation method using risk priority number (RPN) is used in the shipbuilding process. In the case of the conventional RPN method, evaluation items and criteria are ambiguous, and subjective factors such as evaluator's experience and understanding of the system operate a lot on the same contents, resulting in differences in evaluation results. Therefore, this study aims to evaluate the safety and reliability for ship application of the reliability-enhanced fuel cell-based hybrid power system by applying the re-established FMEA technique. Experts formed an FMEA team to redefine reliable assessment criteria for the RPN assessment factors severity (S), occurrence (O), and detection (D). Analyze potential failures of each function of the molten-carbonate fuel cell (MCFC) system, battery system, and diesel engine components of the fuel cell-based hybrid power system set as evaluation targets to redefine the evaluation criteria, and the evaluation criteria were derived by identifying the effects of potential failures. In order to confirm the reliability of the derived criteria, the reliability of individual evaluation items was verified by using the significance probability used in statistics and the coincidence coefficient of Kendall. The evaluation was conducted to the external evaluators using the reestablished evaluation criteria. As a result of analyzing the correspondence according to the results of the evaluation items, the severity was 0.906, the incidence 0.844, and the detection degree 0.861. Improved agreement was obtained, which is a significant result to confirm the reliability of the reestablished evaluation results.

Keywords: hybrid power system; failure mode and effects analysis; risk priority number; ship safety; Kendall's coefficient

1. Introduction

In the early 2000s, the Maritime Safety Committee (MSC) of the International Maritime Organization (IMO) adopted the item goal-based new ship construction standards (GBS) [1], which present new ship design and construction concepts for the long-term organizational work plan. They then developed safety level approach (SLA)-based GBS that are applicable to all ships [2]. The IMO has since actively strengthened the Safety of Life at Sea (SOLAS) standards based on the GBS to reduce the underlying causes of marine accidents and environmental pollution from ship construction and to prioritize ship safety [3].

To assess safety in the ship construction stage, a hazard identification and risk analysis (HIRA) is conducted to identify and evaluate the risk of the system installed in a ship. Specific evaluation methods for analyzing hazards in HIRA include hazard identification (HAZID), hazardous operability (HAZOP), what-if/checklist, and failure mode and effects analysis (FMEA) [4].

FMEA, a type of risk assessment method, was developed for the Apollo project by the National Aeronautics and Space Administration (NASA) in the mid-1960s. Since then, three major US automakers have introduced their own assessment system "QS-9000" [4]. However, FMEA is the most common way to evaluate device reliability [5]. It is a preventive reliability assessment method performed at the design stage for system or component changes, and it uses an empirical perspective for the analysis and component changes to achieve the optimal results. It is extensively used to assess the design, process, and system risks across all industries including the shipbuilding and marine sectors.

FMEA is advantageous in that it enables systematic analysis using simple methods. The evaluation criteria for the expected severity, occurrence, and detection are established using the risk priority number (RPN) technique, and the failures for individual components are assessed [5,6]. These results are combined to obtain the criticality. However, the logic is inferior to other methods because it uses a qualitative evaluation, and the evaluation results may vary depending on the experience or inclination of the evaluator assessing the failure.

Researchers have performed various studies to increase the objectivity of FMEA. Research has been conducted on an approach combining FMEA and the Boolean representation method (BRM) [7], a method that describes the elements required for FMEA and then develops and applies an appropriate FMEA form for an effective evaluation. Studies have also used a computer system method that supports FMEA evaluations on the Internet [8], the risk priority ranks (RPR) approach to prioritize failure modes [9], a method based on fuzzy logic that considers the interdependence between various failure modes [10–14], a fuzzy-based FMEA performance improvement method using GRAY relationship theory [15], and a method that provides a framework for automatically generating FMEA from past FMEA data using functional inference techniques [16]. Research has additionally been conducted on how to most effectively apply the FEMA system due to difficulties related to its numerous subsystems and the lack of consideration for the indirect relationship between the components in the RPN technique.

In particular, in recent years, in order to apply environmentally friendly ships, ships using hybrid fuel cells, batteries, etc. are being operated mainly on small coastal ships. These vessel systems are very different from the diesel engines used as conventional ship power sources, so new FMEA evaluation criteria and items should be provided to evaluate the safety and reliability of such vessels. However, even in shipyards that are currently building vessels, FMEA evaluation criteria or items have not been specifically set.

Therefore, in this study, the proposed FMEA was conducted to secure the safety and reliability for applying the fuel cell-based (molten-carbonate fuel cell (MCFC; 100 kW), battery (30 kW), and diesel generator (50 kW)) test bed to the actual ship. We analyzed various problems in evaluating RPN, which is mainly used in FMEA, and formed an FMEA expert team to select evaluation criteria and items. As a result, we developed a worksheet applying the reestablished RPN evaluation criteria, and applied Kendall's coefficient of correspondence to the existing evaluation results and the reestablished evaluation results for objective determination of the reestablished evaluation criteria. It was confirmed that the reestablished assessment in the FMEA evaluation of the combined power source showed more reliable results. In addition, the criteria for establishing countermeasures based on the results of the FMEA were prepared, and the proposed evaluation method was found to be effective for the application of the assessment of the safety and reliability of the combined power source.

2. Theoretical Background of FMEA and RPN Introduction

2.1. What is FMEA?

FMEA was first used in the NASA Apollo project in the 1960s. In 1974, it was used to develop United States Navy missiles and was established as the United States MIL-STD-1629 standard. Afterwards, the QS-9000 standard was established by the United States automobile industry, and FMEA was introduced in all industries, including shipbuilding [4]. The FMEA method prioritizes resources, ranks risks, and creates an activity and control plan to analyze the target system [5,6], thereby analyzing failure types and their influence and examining improvement measures with consideration of criticality [17].

The objectives of FMEA are as follows:

(1) Identify potential defects inherent in the system and evaluate the severity of their effects.
(2) Identify key management items.
(3) Recognize important potential design and process defects
(4) Prevent severe product accidents and customer complaints.
(5) Provide a basis for establishing sector-specific measures to eliminate or reduce defects.
(6) Enhance efficiency by verifying design and production problems.

Figure 1 illustrates the typical FMEA process.

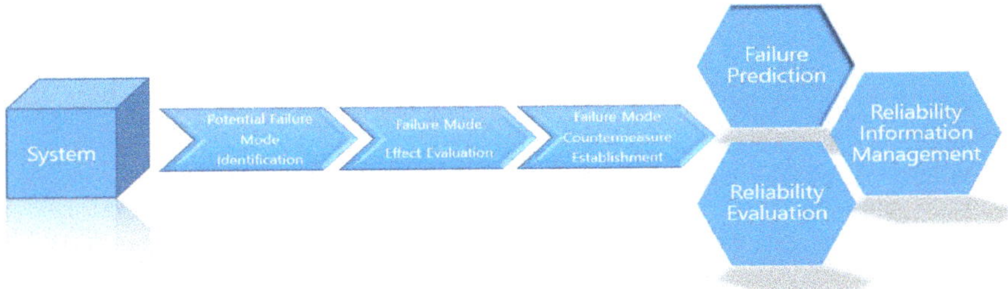

Figure 1. Block diagram of the typical failure mode and effects analysis (FMEA) process.

2.2. RPN Technique

RPN is a relative potential failure evaluation measure that is primarily used in FMEA, and it prioritizes management and corrective actions. RPN consists of three items: severity (S), occurrence (O), and detection (D). The value of each item is divided into 10 levels from 1 (bad) to 10 (good); the value of RPN is between 1 and 1000 and is obtained by the product of each item [17–20]:

$$\text{RPN} = \text{S} \times \text{O} \times \text{D}.$$

Figure 2 shows the meaning of each RPN item. S affects the customer in relation to the process or product when a potential failure occurs. The degree of S of the effect is scored and evaluated, and a reduction in the S class can only be affected through design changes. O refers to the possibility of the cause and mechanism taking place. The O class must be consistent. D indicates whether a potential failure mode and its cause can be discovered or detected.

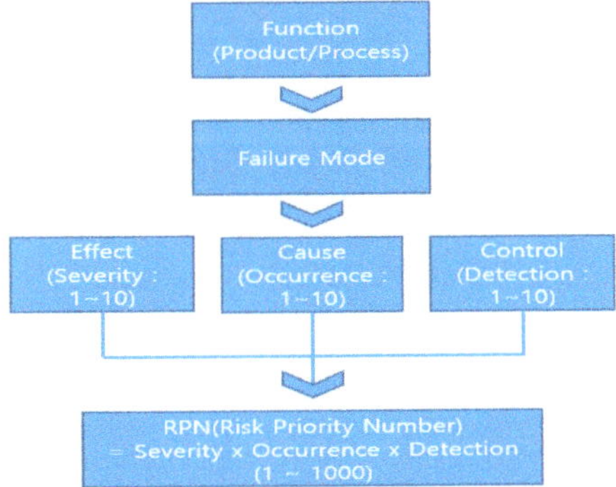

Figure 2. The meaning of each risk priority number (RPN) item.

3. Problem Analysis and Solution of the Existing RPN Evaluation Method

3.1. Problems with the Existing RPN Evaluation Method

There are numerous problems with the existing RPN evaluation method; the following issues directly affect the evaluation [4,18,19].

(1) S, O, and D, the components of RPN evaluation, are influenced by many subjective factors that depend on the evaluator. Therefore, if the evaluator is insufficiently experienced with and knowledgeable of the system, the results may differ from those of another evaluator using the same criteria. The evaluation results of RPN are sensitive to the score variations of each component (S, O, and D). Therefore, if the evaluation criteria are unclear, the evaluation results can differ. For example, assuming that S and O are fixed at a class of 7 and D has a 1 class difference, the RPN score varies by a sizeable 64 points.

(2) In some cases, the evaluation criteria are inappropriate for the particular product or system being evaluated. For example, the RPN standards for shipbuilding differ significantly from those of automakers; applying uniform criteria to both systems greatly increase the likelihood of issues occurring when operating the product.

(3) While the evaluation components of RPN can be assessed individually, the influence of S, O, and D on each other is not taken into account. For example, assume that for RPN1, S, O, and D are 4, 5, and 6, respectively, and the RPN has value of 120. The S, O, and D of RPN2 are 4, 6, and 6, respectively, and the total RPN is 144.

(4) The evaluator responsible for the system is in charge of establishing and implementing measures; therefore, they may be reluctant to thoroughly evaluate the system RPN and may intentionally underestimate it. RPN underestimation and product recalls can lead to enormous time and financial losses, and damage to the manufacturer's image.

(5) If the system evaluation criteria are ambiguous, the evaluator may assess them arbitrarily, leading to vast RPN differences between evaluators.

Overestimating RPN leads to the implementation of unnecessary countermeasures and an excessively safe system design, increasing system installation costs. In contrast, if RPN is underestimated, the appropriate measures for the effects of each failure mode are not established, risking the preventability of future accidents. This can then lead to huge time and money losses. For example, in 1998, GM in the United States received a $4.97 billion fine to compensate the explosion of an automobile fuel tank following a traffic accident. According to the company internal report, the reliability assessment recognized that there was an explosion risk if the fuel tank was manufactured at a low cost. In spite of having access to this information, the vehicle was released without any modifications, leading to the highest payout for individuals in American history [5].

Although FMEA poses numerous problems, it is the most frequently applied reliability evaluation method across all industries because of its simple and systematic analysis. To strengthen the FMEA evaluation performance to supplement the existing problems of FMEA, researchers have investigated methods and approaches from various perspectives [21,22], including a method where, after pre-selecting the factors necessary for FMEA [23], the relationship between the failure mode and effect can be determined by applying various control methods such as fuzzy logic, neural network, functional inference theory [10–14,24–28]; a FMEA matrix, which graphically assesses the relationship between the elements of a system, failure modes, and failure effects [29,30]; methods to effectively prepare the appropriate FMEA form for a given objective [31,32]; methods to provide a worksheet that automatically generates the FMEA using past FMEA data [33]; and other approaches to derive more objective FMEA results [34].

3.2. Improvement in the RPN Technique and Improvement of the Evaluation Method Using Kendall's Concordance Coefficient

To improve the problems that occur in RPN evaluation using FMEA and derive objective results, as shown in Figure 3b, this study precisely identified the potential failure types matching the characteristics of the fuel cell-based hybrid power system for ships and analyzed the RPN evaluation criteria. Figure 3a shows the process for determining the existing RPN evaluation items, and Figure 3b shows the process for determining the RPN evaluation items applied in this study.

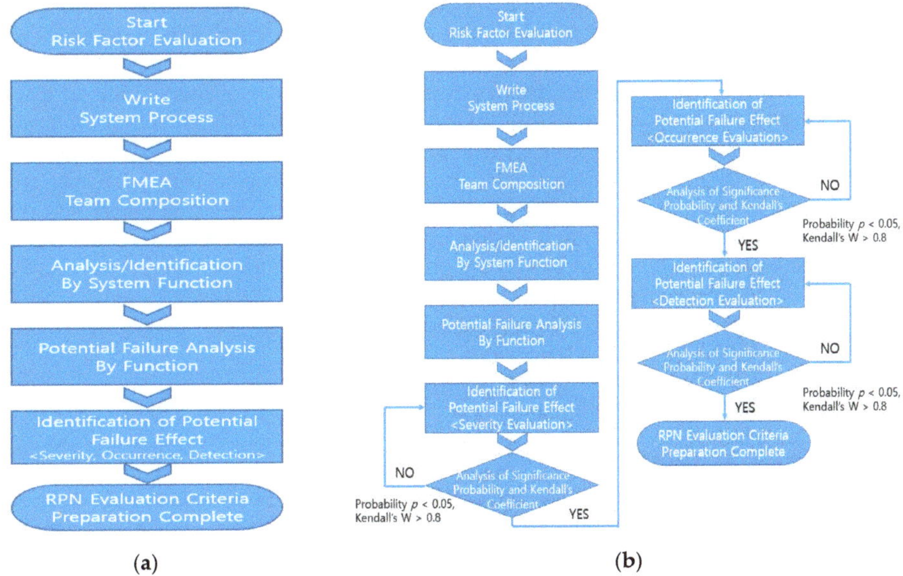

Figure 3. (**a**). The process for determining the existing RPN evaluation items. (**b**). The process for determining the RPN evaluation items applied in this study.

Minimizing differences between the results of various evaluators can increase RPN evaluation reliability. To increase the reliability of the evaluation results, team members with a certain amount of experience in specialty fields were selected for the FMEA team in this study. They performed a system analysis by function. The FMEA team is aware of the problems with existing FMEA because it has been working in the field for a certain period of time and selected experts with basic experience in FMEA evaluation. Therefore, we understand the importance of FMEA evaluation criteria and item setting.

The composition of the FMEA team and the criteria for selecting experts are as follows:

(1) The FMEA team consists of 10 experts for the group;
(2) The selected experts are currently employed in shipyards, research institutes, classification society, engine makers, test and certification institutes, and educational institutions;
(3) Over 5 years of experience in fuel cell, battery, and diesel engine system;
(4) Have more than 10 times of experiences in evaluation FMEA.

Based on the functional analysis of potential failures, this study designed clear evaluation criteria for S, O, and D. The existing effects of potential failures were identified, then the RPN evaluation criteria were created, and an evaluation was immediately performed. However, when creating the RPN evaluation criteria, this study identified the effects of the potential failures of S, O, and D. The reliability of the evaluation criteria were then confirmed, and the criteria were established using the following procedure.

First, the evaluation items for S, O, and D were established, after which the following research hypothesis for the evaluation items was set: "the evaluation scores by item of the evaluators will be similar.". The FMEA team then performed its own internal evaluation, confirming the significance probability results for the reestablished evaluation items and validating the research hypothesis. Next, the team RPN internal evaluation results were compared with Kendall's concordance coefficient to determine the reliability of each evaluation item. In this paper, Kendall's coefficient of consensus mentioned to verify the reliability of the evaluation items is one of the methods used in nonparametric statistics to analyze the relationship between phenomena measured on the sequence scale [35]. Kendall's coincidence coefficient is typically used for attribute agreement analysis, with coefficient values ranging from 0 to 1. The higher the value of the coefficient, the stronger the association. If the coefficient is greater than 0.9, the relevance is considered very high and the high or significant Kendall's coefficient means that the evaluators apply essentially the same standard when evaluating the sample [36]. Applying the same criteria decreases the ambiguity of the evaluation items, removing arbitrariness and encouraging objectivity. Then, the significance probability of the evaluation criteria items and the results of Kendall's concordance coefficient were determined. If the reliability of the evaluation criteria was lower than the threshold, then the process returned to the previous steps to identify the effect of potential failures; once the reliability of the evaluation criteria reached the threshold, the evaluation criteria was confirmed.

This final evaluation criteria were then used as the basis to assess the external evaluators. Finally, by comparing the results with the existing evaluation criteria, this study numerically confirmed the high reliability of the reestablished evaluation criteria.

4. FMEA Methodology of This Study

4.1. FMEA Procedure of This Study

According to the IEC 60812 standard, the FMEA procedure can be divided into three steps: the preparation, performance, and finishing [37].

4.1.1. Preparation Step

To implement FMEA, it is necessary to examine the criteria applied to each power source and hybrid power system configured in the test bed. As a marine fuel cell was applied to an Eidesvik Offshore support vessel of, this study collected and referenced safety-related data such as fuel supply facilities, fire protection facilities, and ventilation systems. This study also examined the "Guidance for Fuel Cell Systems on Board of Ships" published in the Korean Register of Shipping, the "Approval in principle fuel cell installation for LNG Tanker" standard, and the "Guideline for the use of fuel cell systems on board of ships and boats" published in the Det Norske Veritas-Germanischer Lloyd (DNV-GL) registrar [38].

The FMEA worksheet, an important component of FMEA, should be confirmed before performing FMEA. S classification, one of the items in the worksheet, is particularly important; this should be completed with reference to Table 1, which indicates the severity class presented in IMCA M 166 [39].

Table 1. Example of severity ratings as outlined in IMCA M 166.

Classification	Degree	Description
1	Minor	Functional failure of machinery and process components without the effects of injury, damage, or contamination.
2	Critical	Failure without severe damage, contamination, or injury to the system.
3	Major	Critical damage to the system, including the possibility of injury or minor contamination.
4	Catastrophic	Failure causing total system loss with high possibility of fatal injury or large contamination.

4.1.2. Performance Step

In the FMEA performance step, the causes, effects, countermeasures, and severity for each failure mode were discussed; these items were recorded and organized through a worksheet [4]. Here, the effect of the failure mode could be confirmed through the experience of the evaluator, drawings, or simulations. The RPN was used in the evaluation, which indicates the S, O, and D when performing FMEA [5].

4.1.3. Finishing Step

In the FMEA finishing step, FMEA was performed, and all the generated data were organized into a report. The standards, design drawings, single line diagrams, and worksheets used in the report should be organized in a manner that is useful as design data and for the revision step of the system conducted later on. Figure 4 is the FMEA one cycle.

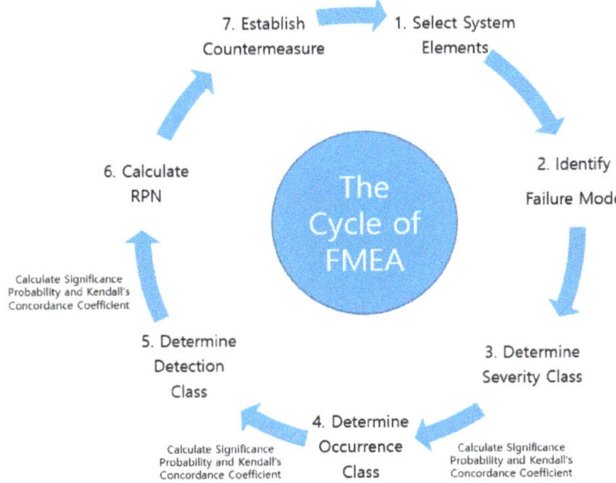

Figure 4. The FMEA one cycle.

4.2. RPN Evaluation Criteria Reestablished in This Study

4.2.1. RPN Evaluation—Severity Criteria

Table 2 shows the S criteria, one of the RPN evaluation factors for a fuel cell-based hybrid power system. The newly applied evaluation criteria were classified as 1, 2, or 3 to enable the accurate evaluation of S from the system and the customer perspectives. Evaluation Criteria 1 simultaneously reflects both the system and customer effects, while Evaluation Criteria 2 contains the corresponding more detailed effects. Evaluation Criteria 3 consists of the effects on the development stage.

Table 2. RPN criteria for severity.

Class	Severity			
	Evaluation Criteria 1		Evaluation Criteria 2	Evaluation Criteria 3
	System Effects	Customer Effects	Detailed Effects	Development Effects
10	With no prior warning, system operations are affected or there are inconsistent international regulations.	There is a major failure related to safety (casualty), such as ignition or explosion without prior warning, posing a risk to customers.	There is a gradual failure after a potential failure related to casualties.	The system development task is dropped.
9	Even with prior warning, system operations are affected or there are inconsistent international regulations.	Even with prior warning, there is a major failure related to safety (casualty), such as ignition or explosion, posing a risk to customers.	There is the sudden occurrence of a dangerous failure directly related to a casualty; items are regulated by the government.	Product concept is changed.
8	The system fails to operate due to the loss of major system functions.	Customers are very dissatisfied, the product does not function, and the product must be disposed of.	Equipment is damaged, it does not operate correctly, and it must be completely disposed of.	There is a change in the assembly component design (customer specifications (spec) out).
7	The system can operate, but the product malfunctions.	Customers are dissatisfied, and the product does not work properly.	While system rework or repair is possible, its functionality has already been severely affected, and selective disposal is required.	There is a change in the component design concept (insufficient customer spec in margin).
6	The main functionality of the system operates normally, but the peripherals are inoperable due to performance deterioration.	Customers are slightly dissatisfied, and simple repair and rework is needed.	Product functionality is affected, and customers are dissatisfied; a partial simple repair is required, and total rework is possible.	There is a change in the component design (internal spec out).
5	The main functionality of the system operates normally, but the peripherals do not operate properly due to performance deterioration.		A repair is required due to degraded product functionality; some functions do not work, and selective rework is possible.	The component is optimized (insufficient spec in, margin).
4	When the system is manufactured, certain peripheral functions are degraded because finishing was not performed properly.	At least half of the customers are mildly dissatisfied; functionality is somewhat affected, but no repair is required.	There is a weak effect on product operations; customers feel discomfort.	Process matching occurs (insufficient spec in, margin).
3		Some of the customers are mildly dissatisfied; functionality is somewhat affected, but no repair is required.	The output or functionality of the unit process is slightly degraded.	There is a slight effect on product characteristics (spec irrelevant).
2	There is almost no effect on the system.	There is almost no effect on customers (next process), and there are no quality defects.	There are no effects on the system, product functionality, and next process.	There is a slight effect on the component characteristics (spec irrelevant).
1			It is difficult to detect a failure, though there is some reluctance.	There is no effect.

4.2.2. RPN Evaluation—O Criteria

Table 3 shows the O criteria, one of the RPN evaluation factors for fuel cell-based hybrid power systems. To precisely evaluate O, the evaluation criteria were classified into 1 (failure occurrence frequency), 2 (possibility of occurrence), 3 (high occurrence rate), and 4 (Cpk value). In the third stage, the high incidence rate was evaluated by applying the PPM(Parts Per Million) index and the Cpk statistical tool was used, which measures the ability of the process to produce output within the required specifications. Cpk represents the capability of the process. If both sides have specifications (upper and lower limits) and the center of the distribution does not match the median of both specifications, bias occurs. In order to prepare and evaluate the incidence criteria of the entire system in detail, evaluation criteria were divided into three stages and four stages. In general, the O is considered good when Cpk is 1.33 or greater for a system or a process. The method for obtaining Cpk is as follows [40].

Table 3. RPN criteria for occurrence.

Class	Occurrence			
	Evaluation Criteria 1	**Evaluation Criteria 2**	**Evaluation Criteria 3**	**Evaluation Criteria 4**
	Failure Occurrence Frequency	**Possibility of Occurrence**	**High Occurrence Rate**	**Cpk Value**
10	Very High relationship	Guaranteed occurrence	1/2 = 500,000 PPM	Less than 0.33
9	High relationship		1/3 = 333,000 PPM	0.33↑
8	Somewhat High relationship	Frequent occurrence	1/8 = 125,000 PPM	0.51↑
7	Lower than high relationship		1/20 = 50,000 PPM	0.67↑
6	Higher than normal relationship	Occasional occurrence	1/80 = 12,500 PPM	0.83↑
5	Normal relationship		1/400 = 2500 PPM	1.00↑
4	Lower than normal relationship		1/2000 = 500 PPM	1.17↑
3	Low relationship	Relatively infrequent occurrence	1/15,000 = 66.67 PPM	1.33↑
2	Very low relationship		1/150,000 = 6.67 PPM	1.50↑
1	Almost no relationship	Almost no occurrence	1 or less/1,500,000 = 0.66 PPM or less	1.67↑

To get the value of Cpk, the capability index Cp is required. Cp is calculated to assess the degree of process capability. Cp can be obtained as Equation (1).

$$\text{Cp} = \frac{USL - LSL}{6\sigma} = \frac{Size\ of\ stanard}{Actual\ Process\ Scatter\ index}. \tag{1}$$

Here, USL: upper specification limit and LSL: lower specification limit.

The value of Cpk can be calculated from the measured data. If there is only an upper limit of the specification, if there is only a lower limit of the specification, it can be divided into a case where both the upper and lower limits of the specification, the calculation formula is as follows (2)–(4).

$$\text{Only upper limit of specification : Cpk} = \frac{USL - \overline{X}}{3\sigma}, \tag{2}$$

$$\text{Only lower limit of specification : Cpk} = \frac{\overline{X} - LSL}{3\sigma}, \tag{3}$$

$$\text{When both upper and lower limit are specified}: \text{Cpk} = (1-\text{k}) \times \text{Cp}, \tag{4}$$

where Cp is the capability index and K is the bias.

K is obtained as follows (5).

$$\text{K} = \frac{\frac{(USL+LSL)}{2} - \overline{\text{X}}}{\frac{(USL-LSL)}{2}}. \tag{5}$$

4.2.3. RPN Evaluation—D Criteria

Table 4 shows the D criteria, one of the RPN evaluation factors for fuel cell-based hybrid power systems. The evaluation criteria were divided into 1 (detectability), 2 (detection difficulty), and 3 (detailed description) to minimize ambiguity and ensure evaluation accuracy.

Table 4. RPN criteria for detection.

<table>
<tr><th rowspan="3">Class</th><th colspan="3">Detection</th></tr>
<tr><th>Evaluation Criteria 1</th><th>Evaluation Criteria 2</th><th>Evaluation Criteria 3</th></tr>
<tr><th>Detectability</th><th>Detection Difficulty</th><th>Detailed Description</th></tr>
<tr><td>10</td><td>Failure (problem) condition completely undetectable.</td><td>Not detectable by known methods.</td><td>No control measures able to detect failure type.</td></tr>
<tr><td>9</td><td>Failure (problem) condition undetectable.</td><td>Detection through indirect, uncertain, or unverified methods.</td><td>Very low detectability according to current system management.</td></tr>
<tr><td>8</td><td rowspan="2">In sensory evaluation, while macrography is possible, failure (problem) condition detection is difficult.</td><td>Detected in customer reliability test.</td><td>Low detectability according to system-wide management.</td></tr>
<tr><td>7</td><td>Detected in internal reliability test.</td><td>Very low likelihood of detection.</td></tr>
<tr><td>6</td><td rowspan="2">Failure (problem) condition normally detected.</td><td>Detected in self-mount test.</td><td>Low likelihood of detection</td></tr>
<tr><td>5</td><td>Detected in mass production test.</td><td>Less than 50% probability of detection.</td></tr>
<tr><td>4</td><td rowspan="2">Failure (problem) condition sufficiently detected.</td><td>Detected in component evaluation.</td><td>Detection probability slightly higher than normal, 50% or more.</td></tr>
<tr><td>3</td><td>Detected in initial sample step.</td><td>Slightly high detectability.</td></tr>
<tr><td>2</td><td rowspan="2">Almost certainly automatically detected during the process.</td><td>Detected in design simulation.</td><td>Very high detectability.</td></tr>
<tr><td>1</td><td>Detected in concept design.</td><td>Certainly detected.</td></tr>
</table>

4.3. Evaluation Method for RPN Evaluation Items Using Kendall's Concordance Coefficient

First, the research hypothesis was established for the RPN evaluation items S, O, and D, and the evaluation items reestablished within the FMEA team were evaluated. Based on the results of the internal evaluation, the significance probability was compared to confirm the validity of the research hypothesis for the evaluation items. The process returned to the potential effect evaluation step if the research hypothesis was rejected. Here, 'P' indicates the significance probability, i.e., the probability that the null hypothesis occurs. The probability that the research hypothesis occurs is set to '1-P'; if the significance probability is less than 5%, then the null hypothesis is rejected, and the research hypothesis is supported. Table 5 shows the null and research hypotheses of this study [35].

Table 5. The null and research hypotheses of this study.

Research hypothesis: The evaluation scores by item of the evaluators will be similar, thus resulting in high reliability.
H_0: The evaluation scores by item of the evaluators will not be similar, thus resulting in low reliability. H_1: The evaluation scores by item of the evaluators will be similar, thus resulting in high reliability.

H_0 is the null hypothesis, which refers to the already established hypothesis. H_1 is the research hypothesis, which negates the null hypothesis; it refers to the method of validating the established research hypothesis.

There are many ways to find correlations, but the most common correlation coefficients are Pearson, Kendall, and Spearman. For the FMEA evaluation items, a non-parametric test was applied instead of a parametric test because an analysis method that directly calculates the probability and statistically tests the data is appropriate regardless of the shape of the population. Pearson is basically used for the correlation analysis, but since it is a parametric test that shows correlations when variables are continuous data, one of the Kendall and Spearman's methods was used to apply nonparametric tests without linear correlation. Spearman generally has higher values than Kendall's correlation coefficient, but is sensitive to deviations and errors in the data. Therefore, Kendall's correlation coefficient was applied in this study because the sample size was small and the data dynamics were large.

The internal evaluation of the FMEA team confirmed the validity of the research hypothesis on the reestablished evaluation items, after which the Kendall's concordance coefficient was compared to determine the reliability of the evaluation items for the individual evaluations. Kendall's concordance coefficient indicates a correlation between multiple evaluators assessing the same sample. The coefficient ranges from 0 to 1, with a higher value indicating stronger correlations. Coefficients above 0.9 are generally considered to indicate very high concordance, meaning that the evaluators apply essentially the same criteria when evaluating the samples, decreasing the ambiguity of the evaluation items, removing evaluation arbitrariness, and encouraging objectivity [36]. If the coefficients for each item deviate from the criteria, the process returns to the potential effect evaluation step.

This study calculated Kendall's concordance coefficient using Equations (6)–(8) and Statistical Package for the Social Sciences (SPSS), a widely used program in statistical analysis. The coefficient was calculated to analyze the concordance between the evaluators for the reestablished S, O, and D evaluation results.

$$W = \left[\frac{12\sum T_i^2}{K^2N(N^2-1)}\right] - \frac{3(N+1)}{N-1}, \tag{6}$$

where T_I is the sum of the classes assigned to each target item by the evaluators, K is the number of evaluators, and N is the number of target items.

The formula for calculating T_i finds the mean (R_i) for the sum of sequence scales.

$$\overline{R_i} = \frac{\sum R_i}{N}. \tag{7}$$

Then, the average deviation T_i for each item can be obtained as follows.

$$T_i = \sum (R_i - \overline{R_i})^2. \tag{8}$$

When establishing the evaluation items, the reliability of the internal evaluation results is verified using the significance probability for the research hypothesis for S, O, and D. The Kendall concordance coefficient was applied to reestablish the evaluation items that satisfy the criteria.

Based on the confirmed evaluation items, the external evaluators were requested to simultaneously evaluate both the existing and reestablished evaluation items. The significance probability and Kendall's concordance coefficient could again be applied to the results of the existing and reestablished evaluation items to judge the application of the same standard. Thus, using the reestablished evaluation items, it is possible to verify that the evaluators are making objective, rather than arbitrary, decisions.

5. System Configuration and Subsystem Classification of the Hybrid Power System for Ships Subject to FMEA Evaluation

The fuel cell-based hybrid power system for ships consists of a power generation, power distribution, output performance verification, and control and management system. Figure 5 shows the subsystems of each system. The power produced by the power generation system is dispersed in the power distribution system, and the product of the output verification system is regulated according to the control and management system commands. The power distribution, output performance verification, and control and management systems have already been applied to all the ships currently under construction; hence, their operation reliability is sufficiently secured, and they were excluded from the FMEA of this study. Out of the subsystems of the hybrid power system, the power generation system (i.e., the failure mode and failure effect of the power source) was evaluated.

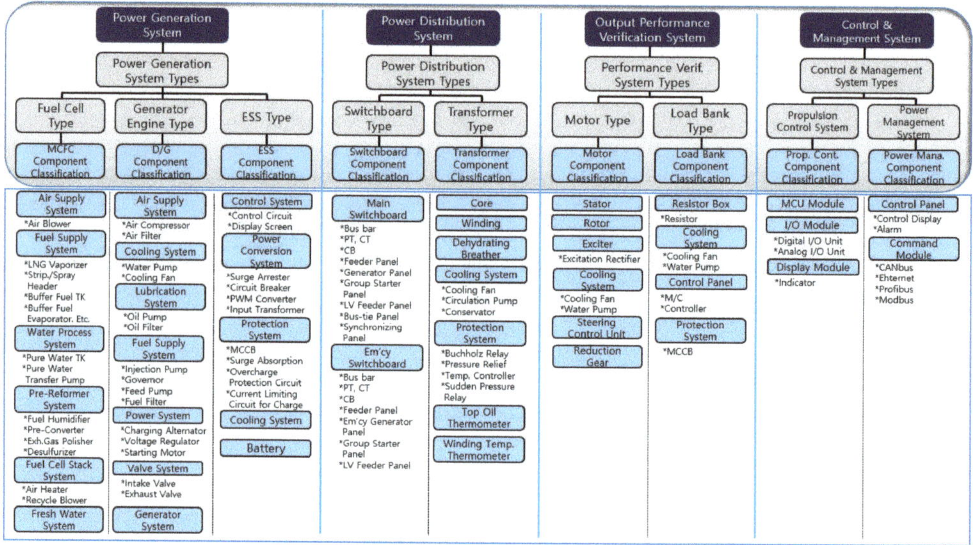

Figure 5. System configuration and subsystem classification of the hybrid power source for the ship.

5.1. Overall Composition of Power Generation System of the Hybrid Power System for Ships

The power generation system can be divided according to the power use purpose, as shown in Figure 6: main power, emergency power, auxiliary power, and alternative maritime power (AMP). The fuel cell, generator, and battery supply power to the main and auxiliary power sources. While AMP is generally supplied from onshore sources through cold ironing, in the hybrid power system, depending on the anchoring period, fuel cells with low greenhouse gas emissions can supply power on board.

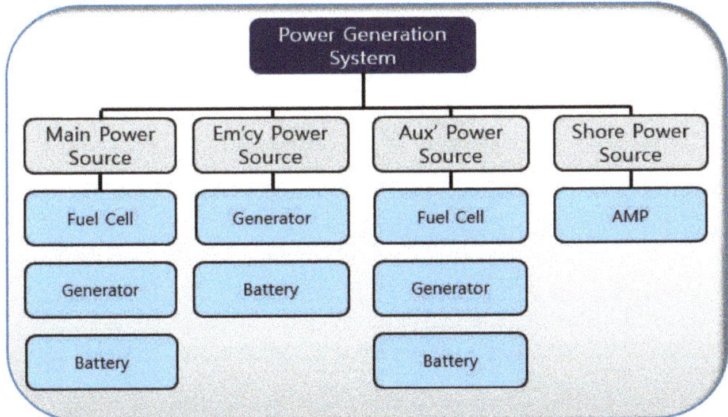

Figure 6. Power generation system of the hybrid power system for a ship.

5.2. Classification of Fuel Cell Components

A molten-carbonate fuel cell (MCFC) generally consists of a regulator, desulfurizer, humidifier, pre-converter, super heater, recycle blower, fresh air blower, inline heater, and catalytic oxidizer [41]. However, MCFCs for ships are comprised of the following components as shown in the block diagram of Figure 7: an air supply system, fuel supply system, water process system, pre-reformer system, fuel cell stack, fresh water system, auxiliary boiler and steam system, and cargo handling system [42].

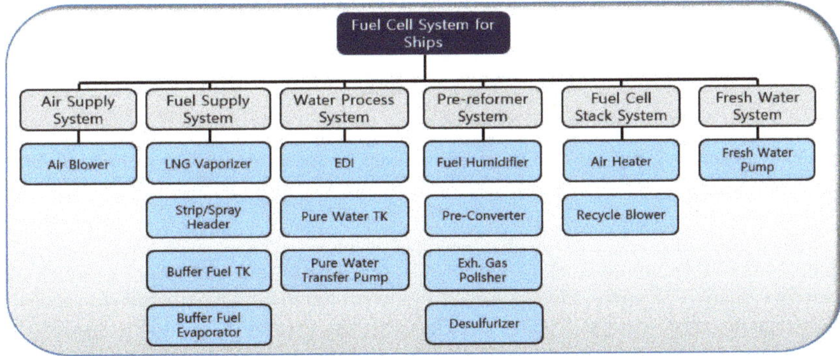

Figure 7. Fuel cell system of the hybrid power system for a ship.

Figure 8 is the configuration of the fuel cell for ships. The MCFC fuel supply system for ships must be connected to the pre-reformer system in the liquified natural gas (LNG) fuel supply chain, the fuel supply system of LNG propulsion ships was selected.

The electrolyte of the molten carbonate fuel cell (MCFC) is alkali metal carbonate, which is a mixture of lithium and potassium or lithium and sodium carbonate contained in a ceramic matrix of $LiAlO_2$. In general, it operates at a high temperature of 600–700 °C and carbonate ions ($CO_3{}^{2-}$) act as a charge carrier. Figure 9 and Equations (9)–(11) show a schematic diagram and chemical reactions occurring in MCFC [41].

$$\text{Total Reaction}: \; H_2 + \frac{1}{2}O_2 + CO_2 \rightarrow H_2O + CO_2. \tag{9}$$

$$\text{Anode Reaction}: \; H_2 + CO_3^{2-} \rightarrow CO_2 + H_2O + 2e^-. \tag{10}$$

$$\text{Cathode Reaction}: \; \frac{1}{2}o_2 + CO_2 + 2e^- \rightarrow CO_3^{2-}. \tag{11}$$

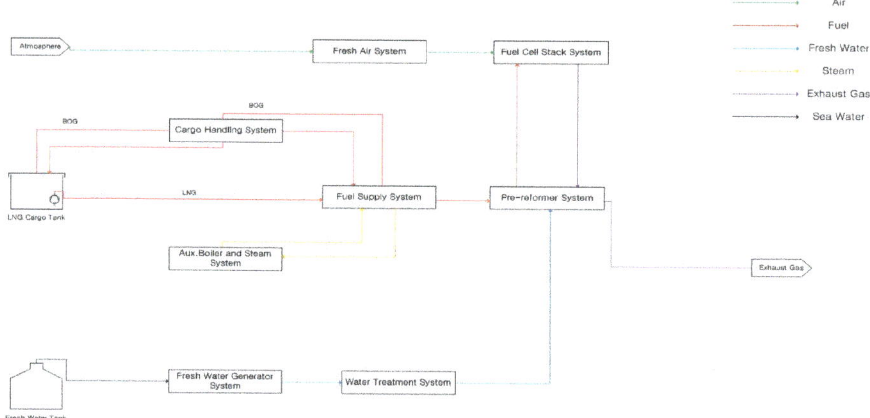

Figure 8. Composition of the fuel cell system for a ship.

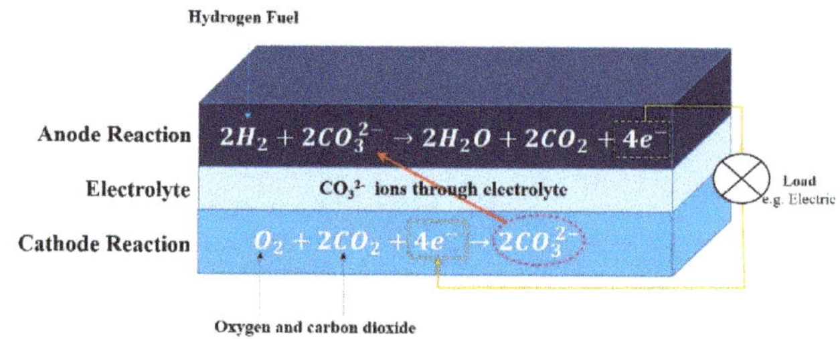

Figure 9. The schematic diagram and chemical reactions for the molten-carbonate fuel cell (MCFC) using hydrogen fuel.

MCFC needs to be supplied carbon dioxide together with oxygen to the cathode. The supplied carbon dioxide is converted into carbonate ions and becomes a means of moving ions between the cathode and the anode. The transferred carbonate ions are converted back to carbon dioxide by reaction with hydrogen at the anode side, and water and electricity are generated together as a result. In MCFC, not only hydrogen but also carbon monoxide can be used as fuel. Figure 10 schematic diagram and chemical reactions for MCFC using carbon monoxide fuel.

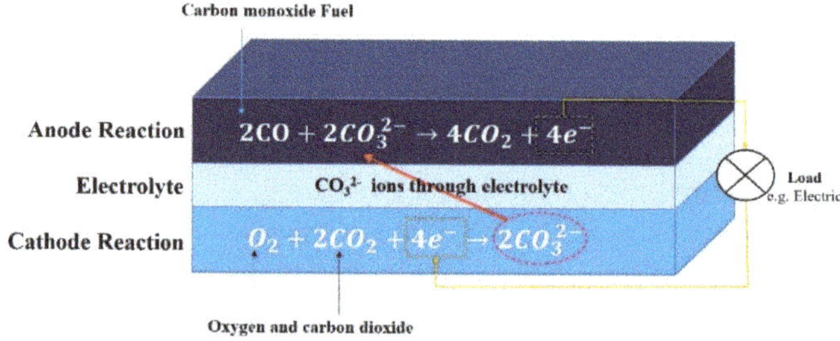

Figure 10. The schematic diagram and chemical reactions for the MCFC using carbon monoxide fuel.

In case of using carbon monoxide as fuel, the chemical reaction of the cathode is the same as that of using hydrogen as fuel. Oxygen and carbon dioxide supplied to the cathode react with each other to be converted to carbonated ions, which are transferred to the anode through the electrolyte. The transferred carbonate reacts with carbon monoxide supplied to the anode side and is converted back to carbon dioxide.

5.3. Classification of Generator Engine Components

The systems of diesel engines, which have been most commonly used as driving generators, include the air supply, cooling, lubrication, fuel supply, power, and valve systems, while electric equipment includes electric governors, measuring equipment, control and safety devices, and cooling devices. In the case of governors in particular, electronic equipment is currently used. It receives the signal from the power management system (PMS) and adjusts the engine speed using torque control. Figure 11 shows the generator engine components.

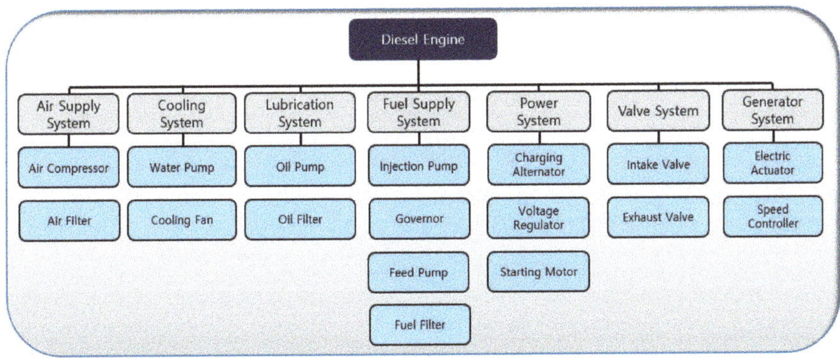

Figure 11. Diesel engine system of the hybrid power system for a ship.

5.4. Classification of Energy Storage System (ESS) Components

An energy storage system (ESS) is divided into control, cooling, protection, and power control systems. The protection system includes a reverse current protection device in the event of power failure, Molded Case Circuit Breaker (MCCB) and surge absorbing element, overcharge protection circuit, and charging current limiting circuit. The representative power conversion systems include pulse width modulation (PWM) converters and input transformers. Figure 12 shows the ESS components.

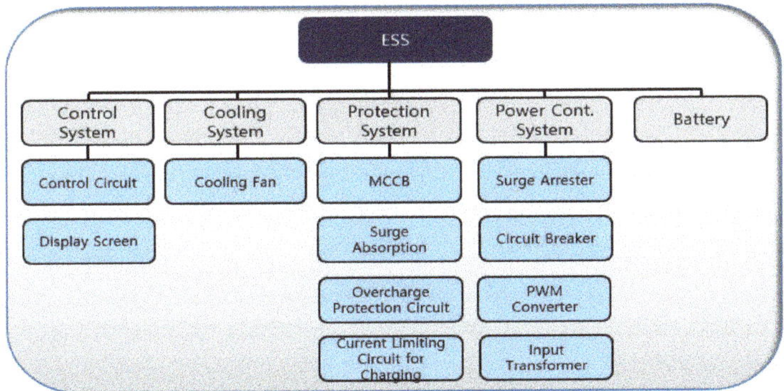

Figure 12. Energy storage system (ESS) of the hybrid power system for a ship.

6. Analysis of FMEA Performance Results

Before applying the fuel cell-based hybrid power system to actual ships, this study first performed FMEA to evaluate the system stability and reliability using onshore test beds. The types of failures that may occur in ship applications were identified, their effects were assessed, and corresponding improvements and supplements to the system were proposed.

The power produced from the hybrid power generation system was distributed through the power distribution system, passed through a synchronization system, and was converted to a voltage and frequency suitable for the output performance verification system. Table 6 shows the selected types of equipment required for a FMEA of the hybrid power system.

Table 6. Equipment list for FMEA of hybrid power system for ship.

Upper System	Group	Subsystem	Subgroup	Equipment
Fuel cellSystem	1	Air supply system	1.1	Air blower
		Water treatment system	1.2	Water pump, Water tank
		Pre-reformer system	1.3	Desulfurizer
Diesel generator system	2	Air supply system	2.1	Air compressor, Air filter
		Cooling system	2.2	Water pump
		Lubrication system	2.3	Oil pump, Oil filter
		Fuel supply system	2.4	Injection pump, Feed pump, Fuel filter
		Power system	2.5	Charging alternator, Starting motor
		Valve system	2.6	Intake valve
ESS system	3	Control system	3.1	Control circuit
		Power conversion system	3.2	Surge arrester
		Protection system	3.3	MCCB
		Cooling system	3.4	Cooling fan

6.1. FMEA Analysis Results of Fuel Cell System

Based on the FMEA results for the fuel cell system, three systems were examined from highest to lowest RPN, the results of which can be found below.

(1) Coating loss occurs due to the rapid ON/OFF desulfurizer cycle, and the desorption amount is reduced. Stack life is improved by replacing the adsorbent.
(2) Coating loss occurs due to the rapid ON/OFF desulfurizer cycle, blocking the back end desulfurizer filter. Stack life is improved by replacing the adsorbent.
(3) Initial power generation of the fuel cell is impossible due to the excessive flow of the air blower. A low air stoic supply is designed for the ignition of the oxidizer.

Even if the sulfur component contained in the natural gas is 0.2 ppm or less, the desulfurization process is required because the activity of the steam reforming catalyst is lowered and the electrode in the MCFC is poisoned, thereby greatly reducing the performance. Desulfurization methods include hydrogen desulfurization (HDS) and the use of absorbents for desulfurization. The method mentioned in this paper is the use of absorbents, which use activated carbon to absorb and remove sulfur. It is coated with a catalyst to enhance the absorption of sulfur. If this coating is not sufficient, the performance of the absorbent may be degraded. Table 7 is the FMEA results of the ship MCFC system [41].

Table 7. The FMEA results of the ship MCFC system.

Item Information			Failure Mode	Failure Effect	Cause of Failure	Severity	Occurrence	Detection	RPN	Recommended Measures
No	Ref. No.	Component								
MCFC System										
1	1.1	Air blower	Insufficient air flow	Oxidizer temperature limit	Insufficient air flow	8	2	4	64	The air flow required for oxidizer operation is calculated, and the pump is selected.
			Insufficient air flow	Air stoic, Resonance generated due to inconsistency	Insufficient air flow	6	6	3	108	A larger supply is designed than the theoretical air flow required for oxidizer operation.
			Insufficient air flow	Incomplete combustion	Insufficient air flow	9	5	3	135	A larger supply is designed than the theoretical air flow required for oxidizer operation.
			Excessive air flow	Initial ignition impossible	Excessive air flow	8	5	4	160	A smaller air stoic supply is designed for the initial ignition of the oxidizer.
			Insufficient air flow	Increased CO concentration in reformate	Insufficient air flow	8	6	3	144	The air flow relative to the amount of reformate that the prox must process is calculated, and the pump is selected.
			Excessive air flow	Pump power consumption increases	Excessive air flow	4	3	3	36	The air flow relative to the amount of reformate that the prox must process is calculated, and the pump is selected.
			Excessive air flow	Prox reaction performance decreases due to fast flow rate, increasing CO	Excessive air flow	8	3	3	72	The air flow relative to the amount of reformate that the prox must process is calculated, and the pump is selected.
			Insufficient air flow	Increased catalytic CO poisoning	Insufficient air flow	8	3	3	72	Select flow rate using the calculated value compared to the amount of hydrogen in the reformate supplied to the stack.
			Excessive air flow	Catalytic oxidation reduces the anode performance and lifetime	Excessive air flow	8	3	3	72	Select flow rate using the calculated value compared to the amount of hydrogen in the reformate supplied to the stack.
			Insufficient air flow	Stack output and lifetime reduction	Insufficient air flow	8	5	4	160	Select flow rate using the calculated value compared to the amount of hydrogen in the reformate supplied to the stack.
			Excessive air flow	Pump power consumption increases	Excessive air flow	4	3	4	48	Select flow rate using the calculated value compared to the amount of hydrogen in the reformate supplied to the stack.

Table 7. *Cont.*

Item Information			Failure Mode	Failure Effect	Cause of Failure	Severity	Occurrence	Detection	RPN	Recommended Measures
No	Ref. No.	Component								
					MCFC System					
	1.2	Water pump	Insufficient water flow	Reformer output reduction	Insufficient water flow	6	6	4	144	Calculate and supply water amount required for reforming.
			Insufficient water flow	Coking generated in reformer	Insufficient water flow	7	6	4	168	Calculate and supply water amount required for reforming.
			Excessive water flow	Degradation of reformer performance due to reduced reformer temperature	Excessive water flow	5	5	4	100	Calculate and supply water amount required for reforming.
	1.3	Desulfurizer	Coating loss	Decrease in sulfur adsorption	Rapid ON/OFF cycle	6	5	7	210	Adsorbent replacement.
				Filter blocked in back end of desulfurizer	Rapid ON/OFF cycle	6	5	7	210	Adsorbent replacement.
			Catalyst crack	Reactor pressure drop increase	Rapid ON/OFF cycle and mechanical shock	8	3	5	120	Adsorbent replacement.
				Decrease in sulfur adsorption	Rapid ON/OFF cycle and mechanical shock	8	3	5	120	Adsorbent replacement.
				Filter blocked in back end of reactor	Rapid ON/OFF cycle and mechanical shock	8	3	5	120	Adsorbent replacement.
			Olefin adsorption	Increase of S concentration in fuel	Excess olefin concentration in fuel	5	3	5	75	Filter replacement.
				Deterioration of filter life	Excess olefin concentration in fuel	5	3	5	75	Filter replacement.
			Water adsorption	Increase of S concentration in fuel	Pre-filter performance deterioration	7	3	5	105	Filter replacement.
				Deterioration of filter life	Pre-filter performance deterioration	7	3	5	105	Filter replacement.
			Catalyst crack	Reactor pressure drop increases	Sudden change in fuel pressure	8	3	5	120	Filter replacement.
				Filter blocked in back end of reactor	Sudden change in fuel pressure	8	3	5	120	Filter replacement.

6.2. FMEA Analysis Results of Diesel Generator System

Based on the FMEA results for the diesel generator system, three systems were examined from highest to lowest RPN, the results of which can be found below.

(1) If the engine power is insufficient due to the inability of the engine to remove impurities in the fuel filter, and the situation persists, engine wear and cracks occur. The fuel filter must be cleaned and replaced frequently to prevent this.
(2) The engine could not be started due to the failure of the starting switch, starting relay, or magnetic kick switch of the starting motor, leading to a dead ship state. To prevent this, the starting motor was disassembled and components were replaced periodically.
(3) Owing to the aging of the air filter, the air intake to the engine was insufficient, and the engine could not be started, leading to a dead ship state. To prevent this, the air filter was frequently cleaned and replaced.

Table 8 is the FMEA results of diesel generator system.

6.3. FMEA Analysis Results of ESS System

Based on the FMEA results for the ESS system, three systems were examined from highest to lowest RPN, the results of which can be found below.

(1) Insulation resistance functionality deteriorated due to soot and metal particles attaching to the MCCB, which might damage the electric equipment at the MCCB back end. In this situation, the MCCB was replaced immediately.
(2) Owing to the control failure of the cooling fan, the electrolyte temperature rose, and the battery capacity was reduced. The ambient temperature should be decreased, and the specific gravity of the electrolyte should be adjusted.
(3) Due to the adjustment failure of the cooling fan, the electrolyte temperature rose, and separator aging and internal short circuiting occurred. To prevent this, the separator should be replaced.

Table 9 is the FMEA results of ESS.

Table 8. The FMEA results of the ship diesel generator system.

Item Information			Failure Mode	Failure Effect	Cause of Failure	Severity	Occurrence	Detection	RPN	Recommended Measures
No	Ref. No.	Component								
Diesel Generator System										
2	2.1	Air compressor	Unable to start engine	Dead ship	Insufficient compression pressure	6	2	3	36	Repair and replace air compressor.
		Air filter	Insufficient engine power	Air supply pump overload	Insufficient air intake	5	4	4	80	Clean or replace air filter.
			Unable to start engine	Dead ship	Insufficient air intake	8	3	5	120	Clean or replace air filter.
	2.2	Water pump	Overheating	Engine wear and tear, cracks	Insufficient coolant transfer	7	2	5	70	Inspect cooling valve or inspect or repair pump. replacement
	2.3	Oil filter	Engine knocking	Cylinder aging	Unable to remove impurities	8	4	3	96	Clean or replace oil filter.
	2.4	Injection pump	Insufficient engine power	Engine cracks	Insufficient fuel injection	7	3	4	84	Adjust injection pump.
			Abnormal idle operation	Injection nozzle cracks	Air intake in injection pump	5	3	4	60	Remove air in pump.
			Excessive fuel consumption	Knocking due to rich burn	Excessive fuel injection	6	3	3	54	Adjust injection pump.
		Feed pump	Insufficient engine power	Knocking	Pump functionality deterioration	6	3	5	90	Repair or replace pump.
		Fuel filter	Insufficient engine power	Engine wear and tear, cracks	Unable to remove impurities	7	4	5	140	Clean or replace fuel filter.

Table 8. *Cont.*

Item Information			Failure Mode	Failure Effect	Cause of Failure	Severity	Occurrence	Detection	RPN	Recommended Measures
No	Ref. No.	Component								
					Diesel Generator System					
	2.5	Charging alternator	Unable to start engine	Dead ship	Electrical wiring slack and short circuit	8	2	3	48	Retighten or replace charging alternator.
		Starting Motor	Unable to start engine	Dead ship	Starting switch failure, starting relay failure, magnetic switch failure	8	3	5	120	Disassemble starting motor.
	2.6	Intake valve	Insufficient engine power	Engine wear and tear, cracks	Incorrect valve clearance	5	5	3	75	Adjust intake valve.
				Engine wear and tear, cracks	Poor valve adhesion	6	4	3	72	Repair intake valve.
			Insufficient compression pressure	Trouble starting	Poor valve closure	7	2	5	70	Analyze fuel injection timing and replace valve.
					Valve spring damage	7	4	3	84	Replace valve spring.

Table 9. The FMEA results of the ESS.

Item Information			Failure Mode	Failure Effect	Cause of Failure	Severity	Occurrence	Detection	RPN	Recommended Measures
No.	Ref. No.	Component								
ESS System										
3	3.1	Control circuit	Control failure	Electrolyte leakage and reduction	Overcharge	8	2	5	80	Replenish purified water and adjust electrolyte volume and specific gravity.
			Control failure	Pole plate corrosion	Overcharge	8	2	4	64	Control floating charge voltage or shorten equalization time.
			Control failure	Pole plate bending and active material drop	Overcharge	8	2	5	80	Inspect charging current and ambient temperature.
	3.2	Surge arrester	Electrical breakdown	PCS function loss	Internal short circuit	7	2	5	70	Replace surge arrester.
	3.3	MCCB	Insulation resistance deterioration	Damage to back end electric equipment	Overcurrent blocking soot, metal particle adhesion	8	2	7	112	Immediately replace MCCB.
	3.4	Cooling fan	Fan rpm adjustment failure	Internal short circuit due to separator aging	Electrolyte temperature increase	7	3	5	105	Replace separator.
			Fan rpm adjustment failure	Foaming at full charge	Electrolyte temperature increase	7	2	4	56	Control floating charge voltage or shorten equalization time.
			Fan rpm adjustment failure	Battery capacity reduction	Electrolyte temperature increase	3	6	6	108	Improve ambient temperature and adjust electrolyte specific gravity.
			Fan rpm adjustment failure	Pole plate bending and active material drop	Electrolyte temperature increase	5	3	4	60	Improve ambient temperature and adjust electrolyte specific gravity.

6.4. FMEA Results for Each System

This study precisely identified the hybrid power system failure types and applied the reestablished RPN criteria to analyze the potential effects of failure. This study sought to derive consistent results between evaluators through newly applied evaluation criteria, obtaining results that could confirm safety and reliability when applied to the hybrid power system of a ship. Before applying Kendall's concordance coefficient, the hypothesis "The evaluation scores by item of the evaluators will be similar" was established according to the research objective. The significance probability between the existing and reestablished evaluation items was compared, confirming the validity of the research hypothesis. Kendall's concordance coefficient was applied using SPSS to confirm the concordance rate of the evaluation results between the existing and reestablished evaluation items as shown in Tables 10–12.

Table 10. Comparison of severity evaluation results between existing and reestablished evaluation items.

Test Statistics of Existing Evaluation Items		Test Statistics of Reestablished Evaluation Items	
K	3	K	3
N	50	N	50
Approximate significance probability	0.000	Approximate significance probability	0.000
Kendall's W	0.700	Kendall's W	0.906

Table 11. Comparison of occurrence evaluation results between existing and reestablished evaluation items.

Test Statistics of Existing Evaluation Items		Test Statistics of Reestablished Evaluation Items	
K	3	K	3
N	50	N	50
Approximate significance probability	0.000	Approximate significance probability	0.000
Kendall's W	0.703	Kendall's W	0.844

Table 12. Comparison of detection evaluation results between existing and reestablished evaluation items.

Test Statistics of Existing Evaluation Items		Test Statistics of Reestablished Evaluation Items	
K	3	K	3
N	50	N	50
Approximate significance probability	0.002	Approximate significance probability	0.000
Kendall's W	0.565	Kendall's W	0.861

In this study, external evaluators assessed the same samples; based on the significance probability for the evaluation results of each item, the research hypothesis was supported. In addition, among the RPN items, the Kendall's concordance coefficient was 0.906 for S, 0.844 for O, and 0.861 for D. Compared to the existing evaluation items, the results for the reestablished evaluation items indicated that each evaluator applied essentially the same criteria when assessing the samples. The reliability of the evaluation results was therefore verified, and criteria for providing countermeasures for each failure mode were established based on the detected results.

To establish the criteria for countermeasures according to the RPN results of the fuel cell-based hybrid power system, it must be decided whether the absolute or relative RPN values will be used as the standard. To establish countermeasures via relative RPN values, the conditions of the targets for comparison must be similar (e.g., the number of items and the content of each item). However, as the internal device configurations and characteristics differ for each system, the number of evaluation items and the type and contents of each item also differ, making it difficult to apply relative criteria.

Therefore, this study defined the criteria of the reestablished evaluation items for countermeasures using absolute RPN values; specifically, the RPN evaluation class was defined as 1–10, and $1 \leq \text{RPN} \leq$

1000. The following were set as the criteria for establishing countermeasures assuming a reliability of 90%: RPN of 100 or more, and either S, O, or D was 8 or more. Table 13 shows the number of items that should be set for each system according to the criteria.

Table 13. The ration of items required to establish countermeasures for a system based on these criteria.

System	Total Items	Number of Failure Modes to Establish Countermeasures	
MCFC	25	RPN 100 or more	17
		8 or more for each	13
Diesel Generator	16	RPN 100 or more	3
		8 or more for each	4
ESS	9	RPN 100 or more	3
		8 or more for each	4
Overall hybrid power system	50	-	-

7. Conclusions

This study conducted a FMEA to evaluate the safety and reliability of a fuel cell-based hybrid power system for ships. Unlike diesel engines that are mainly used as propulsion power sources in conventional ships, new FMEA evaluation criteria and items are needed to apply fuel cell-based hybrid power sources to ships. In the RPN evaluation currently applied to shipbuilding in shipyards, existing RPN evaluations, the evaluation items and criteria are vaguely established; therefore, results for the same evaluation would differ vastly between evaluators. Accordingly, for the FMEA of this study, the evaluation was performed using several external evaluators who applied reestablished evaluation criteria that mitigate RPN evaluation problems. To analyze the concordance of the reestablished evaluation items, a research hypothesis was established, and the significance probabilities and Kendall's concordance coefficient were calculated using SPSS. The concordance coefficient was 0.906 for S, 0.844 for O, and 0.861 for D. The results indicate that each evaluator applied essentially the same criteria when evaluating the samples, demonstrating that the reliability of the evaluation results was high. The criteria used to establish countermeasures for each failure mode were set based on the D results of the evaluation.

Although having the same evaluation configuration for each hybrid power system is essential to establish countermeasures, each system contains different devices and characteristics, therefore, the number and type of evaluation items also differ. Since it is difficult to apply relative criteria, this study instead used absolute RPN values to set the criteria for establishing countermeasures: a RPN of 100 or more and an S, O, or D of 8 or more.

For the FMEA of this study the power generation system of the hybrid power system (i.e., the failure mode and failure effect of the power source) was evaluated. However, future research must conduct FMEA for the entire set of systems including the power generation, power distribution, output performance verification, and control and management systems of hybrid power systems. Future studies must also perform FMEA for different system operation modes (e.g., single and hybrid operation) to identify hazards that may arise in the systems of actual ships during operation. However, in spite of these limitations, the results of this study showed significant results as an evaluation to confirm the stability and reliability for applying a fuel–cell based hybrid power source to several ships.

Author Contributions: Conceptualization, J.K. and H.J.; Methodology, J.K., H.J. and K.P.; Formal analysis, H.J.; Software, J.K. and H.J.; Writing-original draft preparation, H.J.; Writing-review and editing, J.K., H.J. and K.P. All authors have read and agreed to the published version of the manuscript.

Funding: This research received no external funding.

Conflicts of Interest: The authors declare no conflict of interest.

References

1. IMO. Maritime Safety Committee (MSC) 101 Session. Available online: http://bitly.kr/U7R2WN6 (accessed on 22 November 2019).
2. KEIT. Trends in IMO for Ship Safety and Marine Environment Protection. Available online: http://bitly.kr/WgqktaA4 (accessed on 1 November 2019).
3. KMI. IMO International Maritime Policy Trend. Available online: http://bitly.kr/NuQogc3i (accessed on 18 November 2019).
4. Broadribb, M.P. *Guidelines for Integrating Process Safety into Engineering Projects*; Wiely: Hoboken, NJ, USA, 2019; pp. 386–399.
5. Bertsche, B. *Reliability in Automotive and Mechanical Engineering*; Springer: Berlin, Germany, 2008; pp. 96–106.
6. Ciani, L.; Guidi, G.; Patrizi, G. A Critical Comparison of Alternative Risk Priority Numbers in Failure Modes, Effects, and Criticality Analysis. *IEEE Access* **2019**, *7*, 92398–92409. [CrossRef]
7. Wang, J.; Ruxton, C.; Labrie, C.R. Design for safety of engineering systems with multiple failure state variables. *Reliab. Eng. Syst. Saf.* **1995**, *50*, 271–284. [CrossRef]
8. Huang, G.Q.; Nie, M.; Mak, K.L. Web-based failure mode and effect analysis (FMEA). *Comput. Ind. Eng.* **1999**, *37*, 177–180. [CrossRef]
9. Sankar, R.N.; Prabhu, B. Modified approach for prioritization of failures in a system failure mode and effects anlaysis. *Int. J. Qual. Reliab. Manag.* **2001**, *18*, 324–336. [CrossRef]
10. Mohammadi, A.; Tavakolan, M. Construction Project Risk Assessment Using Combined Fuzzy and FMEA. In Proceedings of the 2013 Joint IFSA World Congress and NAFIPS Annual Meeting (IFSA/NAFIPS), Edmonton, AB, Canada, 24–28 June 2013; pp. 232–237.
11. Gan, L.; Pang, Y.; Liao, Q.; Xiao, N.C.; Huang, H.Z. Fuzzy Criticality Assessment of FMECA for the SADA Based on Modified FWGM Algorithm & Centroid Deffuzzification. In Proceedings of the 2011 International Conference on Quality, Reliability, Risk, Maintenance, and Safety Engineering, Xi'an, China, 17–19 June 2011; pp. 195–202.
12. Ahn, J.; Noh, Y.; Park, S.H.; Choi, B.I.; Chang, D. Fuzzy-Based Failure Mode and Effect Analysis (FMEA) of a Hybrid Molten Carbonate Fuel Cell (MCFC) and Gas Turbine System for Marine Propulsion. *J. Power Sour.* **2017**, *364*, 226–233. [CrossRef]
13. Rabbi, M.F. Assessment of Fuzzy Failure Mode and Effect Analysis (FMEA) for Reach Stacker Crane (RST): A Case Study. *Int. J. Res. Ind. Eng.* **2018**, *7*, 336–348.
14. Xu, K.; Tang, L.C.; Xie, M.; Ho, S.L.; Zhu, M.L. Fuzzy assessment of FMEA for engine systems. *Reliab. Eng. Syst. Saf.* **2002**, *75*, 17–29. [CrossRef]
15. Pillay, A.; Wang, J. Modified failure mode and effects analysis using approximate reasoning. *Reliab. Eng. Syst. Saf.* **2003**, *79*, 69–85. [CrossRef]
16. Teoh, P.C.; Case, K. Failure modes and effects analysis through knowledge modeling. *J. Mater. Process. Technol.* **2004**, *154*, 253–260. [CrossRef]
17. Lo, H.W.; Liou, J.J.H. A Novel Multiple-Criteria Decision-Making-Based FMEA Model for Risk Assessment. *Appl. Soft Comput. J.* **2018**, *73*, 684–696. [CrossRef]
18. Bolbot, V.; Theotokatos, G.; Blulougouris, E.; Vassalos, D. Comparison of diesel-electric with hybrid-electric propulsion system safety using System-Theoretic Process Analysis. In Proceedings of the Power & Propulsion Alternatives for Ships, London, UK, 22–23 January 2019; pp. 1–7.
19. Shrestha, S.M.; Mallineni, J.K.; Yedidi, K.R.; Knisely, B.; Tatapudi, S.; Kuitche, J.; Tamizhmani, G. Determination of Dominant Failure Modes Using FMECA on the Field Deployed c-Si Modules under Hot-Dry Desert Climate. *IEEE J. Photovolt.* **2015**, *5*, 174–182. [CrossRef]
20. Certa, A.; Enea, M.; Galante, G.M.; La Fata, C.M. ELECTRE TRI-Based Approach to the Failure Modes Classification on the Basis of Risk Parameters: An Alternative to the Risk Priority Number. *Comput. Ind. Eng.* **2017**, *108*, 100–110. [CrossRef]
21. Zhao, Y.; Fu, G.; Wan, B.; Pei, C. An Improved Cost-Based Method of Risk Priority Number. In Proceedings of the IEEE 2012 Prognostics and System Health Management Conference (PHM-2012 Beijing), Beijing, China, 23–25 May 2012; pp. 1–4.

22. Wu, J.; Tian, J.; Zhao, T. Failure Mode Prioritization by Improved RPN Calculation Method. In Proceedings of the 2014 Reliability and Maintainability Symposium, Colorado Springs, CO, USA, 27–30 January 2014; pp. 1–6.
23. Giannetti, C.; Ransing, M.R.; Ransing, R.S.; Bould, D.C.; Gethin, D.T.; Sienz, J. Product Specific Process Knowledge Discovery Using Co-Linearity Index and Penalty Functions to Support Process FMEA in the Steel Industry. In Proceedings of the 44th International Conference Computer Industrial Engineering, Istanbul, Turkey, 14–16 October 2014; pp. 305–319.
24. Arabian-Hoseynabadi, H.; Oraee, H.; Tavner, P.J. Failure Modes and Effects Analysis (FMEA) for Wind Turbines. *Int. J. Electr. Power Energy Syst.* **2010**, *32*, 817–824. [CrossRef]
25. Kim, D.H.; Kim, S.C.; Park, J.S.; Kim, E.J.; Kim, E.S. Analysis of Risk Priority Number for Grid-Connected Energy Storage System. *J. Korean Soc. Saf.* **2016**, *31*, 10–17. [CrossRef]
26. Mzougui, I.; El Felsoufi, Z. Proposition of a Modified FMEA to Improve Reliability of Product. *Procedia CIRP* **2019**, *84*, 1003–1009. [CrossRef]
27. Schlasza, C.; Ostertag, P.; Chrenko, D.; Kriesten, R.; Bouquain, D. Review on the Aging Mechanisms in Li-Ion Batteries for Electric Vehicles Based on the FMEA Method. In Proceedings of the 2014 IEEE Transportation Electrification Conference and Expo (ITEC), Dearborn, MI, USA, 15–18 June 2014; pp. 1–6.
28. Niu, Y.M.; He, Y.Z.; Li, J.H.; Zhao, X.J. The Optimization of RPN Criticality Analysis Method in FMECA. In Proceedings of the 2009 International Conference on Apperceiving Computing and Intelligence Analysis, Chengdu, China, 23–25 October 2009; pp. 166–170.
29. Lee, W.Y.; Park, G.G.; Sohn, Y.G.; Kim, S.G.; Kim, M.J. Fault Detection and Diagnosis Methods for Polymer Electrolyte Fuel Cell System. *Trans. Korean Hydrog. New Energy Soc.* **2017**, *28*, 252–272.
30. Park, Y.M. A Study on the Application of FMEA for Shipbuilding Process. Master's Thesis, University of Ulsan, Ulsan, Korea, 2008.
31. Youn, J.S.; Park, Y.G.; Um, J.H. Analysis the Railway Accident, Failure through FMEA Technique and Present the Priority Safety Management Items. In Proceedings of the Autumn Conference of the Korean Socity for Railway, Daegu, Korea, 7–9 November 2013; pp. 108–117.
32. Basu, J.B. Failure Modes and Effects Analysis (FMEA) of a Rooftop PV System. *Int. J. Sci. Eng. Res.* **2015**, *3*, 51–55.
33. Zeng, S.X.; Tam, C.M.; Tam, V.W.Y. Integrating Safety, Environmental and Quality Risks for Project Management Using a FMEA Method. *Eng. Econ.* **2010**, *21*, 44–52.
34. Sellappan, N.; Palanikumar, K. Modified Prioritization Methodology for Risk Priority Number in Failure Mode and Effects Analysis. *Int. J. Appl. Sci. Technol.* **2013**, *3*, 27–36.
35. Bao, Z. Tracy-Widom limit for Kendall's tau. *Ann. Stat.* **2019**, *47*, 3504–3532. [CrossRef]
36. Statistics Solutions. Correlation (Pearson, Kendall, Spearman). Available online: http://bitly.kr/smNl299P (accessed on 1 December 2019).
37. IEC 60812. Analysis Techniques for System Reliablility Procedure for Failure Mode and Effects Analysis (FMEA). Available online: http://bitly.kr/DlQDoqSw (accessed on 30 November 2019).
38. DNVGL. Failure Mode and Effect Analysis (FMEA) of Redundant Systems. Available online: http://bitly.kr/7eSXMyz (accessed on 3 December 2019).
39. IMCA. Guidance on Failure Modes and Effects Analysis (FMEA). Available online: http://bitly.kr/TZKtc45x (accessed on 2 December 2019).
40. Process Capability Index. Available online: https://www.whatissixsigma.net/process-capability-index-cpk/ (accessed on 14 January 2020).
41. Larminie, J.; Dicks, A. *Fuel Cell Systems Explained*; Wiely: Hoboken, NJ, USA, 2003; pp. 187–241.
42. Zhang, X.; Higier, A.; Zhang, X.; Liu, H. Experimental studies of effect of land width in PEM fuel cells with serpentine flow field and carbon cloth. *Energies* **2019**, *12*, 471. [CrossRef]

Journal of
Marine Science and Engineering

Article

Life Cycle Performance Assessment Tool Development and Application with a Focus on Maintenance Aspects

Paola Gualeni [1,*], Giordano Flore [1], Matteo Maggioncalda [2] and Giorgia Marsano [1]

1 Department of Naval Architecture, Electrical, Electronics and Telecommunication Engineering, DITEN, University of Genoa, 16145 Genoa, Italy
2 Fincantieri S.p.A. Naval Services, Sales-Integrated Logistic & In Service Support, 16129 Genoa, Italy
* Correspondence: paola.gualeni@unige.it

Received: 15 July 2019; Accepted: 13 August 2019; Published: 19 August 2019

Abstract: Ships are among the most complex systems in the world. The always increasing interest in environmental aspects, the evolution of technologies and the introduction of new rule constraints in the maritime field have compelled the innovation of the ship design approach. At an early design stage, there is the need to compare different design solutions, also in terms of environmental performance, building and operative costs over the whole ship life cycle. In this context, the Life Cycle Performance Assessment (LCPA) tool allows an integrated design approach merging the evaluation of both costs and environmental performances on a comparative basis, among different design solutions. Starting from the first tool release, this work aims to focus on the maintenance of the propulsion system, developing a flexible calculation method for maintenance costs prediction, based on the ship operational profiles and the selected technical solution. After the improvement, the whole LCPA tool has been applied on a research vessel to evaluate, among different propulsion layout solutions, the one with the more advantageous performance in terms of costs during the whole vessel operating life. The identification of the best design solution is strictly dependent on the selection criterion and the point of view of the interested parties using the LCPA tool, e.g., the shipbuilder or the ship-owner.

Keywords: life cycle; maintenance costs; propulsion system maintenance; research vessel

---

1. Introduction

The increasing complexity of ships, the need of better performance and the introduction of new regulations, for example to protect the environment, call for an evolution of the design approach in the maritime field as well. The primary need to remain competitive in the international shipping arena concurs to develop an integrated approach, with superior attention during the design process to economic and environmental aspects along with the ship operational life [1–4]. This new awareness inevitably calls in turn for intensive design comparisons and trade-off analysis with a specific focus on the energy system. In this context also, the comparison between traditional and alternatives fuels plays an important role [5]. The new approach includes market and demand analysis, integrated economic and environmental considerations, efficiency and performance assessment mutually conversant in the ship design process. As possibly useful support in this direction, a formulation of a rational parametric analysis of the vessel over its life cycle has been developed and implemented in the traditional ship design process. In this way, different ship configurations or operational profiles can be deeply analyzed at an early design stage, when designers can still make decisions to change and improve the project [6].

The HOLISHIP European project (www.holiship.eu) addresses the issue of a comprehensive approach for ship design capable of meeting the future market needs. As part of the wide HOLISHIP project, an LCPA (Life Cycle Performance Assessment) tool [7], that combines the LCC (Life Cycle

Costing) and LCA (Life Cycle Assessment), has been developed: the design tool permits, by a comparison framework, the valuation of various design alternatives. It allows a cost/benefit assessment over the ship life cycle considering, at the same phase, both design and operations. Different ship configurations or system layouts can be compared and analyzed in terms of building and operational cost, as well as environmental impact [8].

This paper aims to further develop operational and maintenance costs, defining, in particular, a structured and flexible tool to evaluate maintenance costs for different ship solutions. Based on real maintenance plans and maintenance tasks, overcoming the empiric formulations, this part would be implemented as an element of the main LCPA tool.

Section 2 of this paper provides a general overview of the LCPA tool developed during the HOLISHIP project; Section 3 describes the maintenance techniques commonly used; Section 4 describes step by step the maintenance prediction model developed during the research project; Section 5 provides a complete description of the reference vessel used for the application exercise; Section 6 contains the process to discuss about preferable solutions; Section 7 describes the obtained results; Section 8 provides the conclusions.

2. The LCPA Tool

The acronym LCPA appeared for the first time some years ago on the scenario of EU funded project (BEST, JOULES, RAMSES, and more recently SHIPLYS) since a fully compliant LCA procedure is simply not applicable to ships because of their complexity.

In HOLISHIP, evolution has been proposed, with a specific focus on the possibility to obtain a single and comprehensive index to characterize the ship performance, inclusive of both economic and environmental terms.

The LCPA tool is an integrated software to allow for a life-cycle ships assessment during the design phase. LCPA efficiently merges the LCC and LCA performances that, therefore, can be observed comprehensively from the perspective of the designer/ship-builder, as well as of the ship-owner. The calculation of selected Key Performance Indicators (KPIs) enables the evaluation of the LCPA Index that describes the performance of a baseline ship compared to other alternative design configurations [7,9]. The overall index is a linear combination of selected KPIs that have been properly normalized. The relevant formulations are reported in the following Equations (1)–(3), but further details can be found in [7,10]. In principle, the methodology is open for the integration of further aspects, besides economic and environmental ones, e.g., safety performances [11], in case adequate KPIs are identified.

$$I_{LCC} = \sum_{i=1}^{N_{LCC}} f_{i,LCC} * c_{i,LCC} \leq 1 \; ; \; where : \sum_{i=1}^{N_{LCC}} f_{i,LCC} = 1 \tag{1}$$

$$I_{LCA} = \sum_{i=1}^{N_{LCA}} f_{i,LCA} * c_{i,LCA} \leq 1 \; ; \; where : \sum_{i=1}^{N_{LCA}} f_{i,LCA} = 1 \tag{2}$$

$$I_{LCPA} = f_{LCC} * I_{LCC} + f_{LCA} * I_{LCA} \; ; \; where \; f_{LCC} + f_{LCA} = 1 \tag{3}$$

It is worthwhile mentioning that different approaches are possible to combine economic and environmental aspects, for example, converting all incomparable values into monetary values [12]. To better understand the LCPA software and its further improvements developed in this paper, a synthesis of the main topics are proposed in the following sections. As already mentioned, Key Performance Indicators (KPIs) are the base of the LCPA tool. They are performance indicators resulting after the definition of the problem and its cost drivers. There are, respectively, economics and environmental KPIs that play a fundamental role in the decision-making process:

- **Building Cost (BLD)**: It is the costs sustained by the shipyard to build the vessel. Different estimation techniques can be adopted to evaluate this figure, based on the level of details already known in the actual ship design phase. [13].
- **Capital Expenditure (CAPEX)**: It is the funds that a shipowner use to purchase a vessel from a shipyard. This cost is influenced by the type of ship and the installed technological level. It is directly related to the shipyard Building Cost, but it is also influenced by the market situation at a specific moment.
- **Operating Expenditure (OPEX)**: It is the costs related to the vessel normal operation activities [14]. The OPEX changes during the lifetime of the ship, so it is important to determine how much it changes every year (on average), without considering the change of money value due to the discount rate. OPEX takes into account the operating costs, the voyage costs, the costs related to the payload, insurance, interest, and maintenance, and repair costs.
- **Maintenance and Repair costs (MandR costs)**: They are a part of OPEXs, but they can also be used independently to assess costs related to maintenance of systems and structures. This value is directly influenced by the type of systems and maintenance approach adopted.
- **Net Present Value (NPV)**: It is probably the most popular economic measure. It is an index on the profitability, which is evaluated by subtracting the present values of cash outflows (including initial cost) from the present values of cash inflows over time [15].
- **Energy Efficiency Design Index (EEDI)**: It is a parameter introduced by the Marine Environment Protection Committee (MEPC) of IMO in 2011 and it indicates the energy efficiency of a ship in terms of generated CO2 (grams or tons per mile cargo carried). It is calculated for a specific reference ship operational condition. The aim is to drive ship technologies development to more energy-efficient ones [16].
- **NOx** and **SOx** emissions (during operation): These parameters measure the grams of NOx or SOx generated per unit of transport work. The regulatory framework is imposing, in recent years, stricter and stricter criteria depending on the sea area where the ship is sailing. NOx emissions mainly depend on the type of engines installed on board, while SOx emissions are mainly related to the type of fuel used to produce energy [17].

Further considerations about the significance of such KPIs can be found in [18]. explicit equations for estimation of all these values, are given in [7,10].

The relations between these KPIs are provided in Figure 1a–c. It is important to denote that the CAPEX value has been described in the paper for the sake of completeness, but during this work, we have always referred to the BLD value.

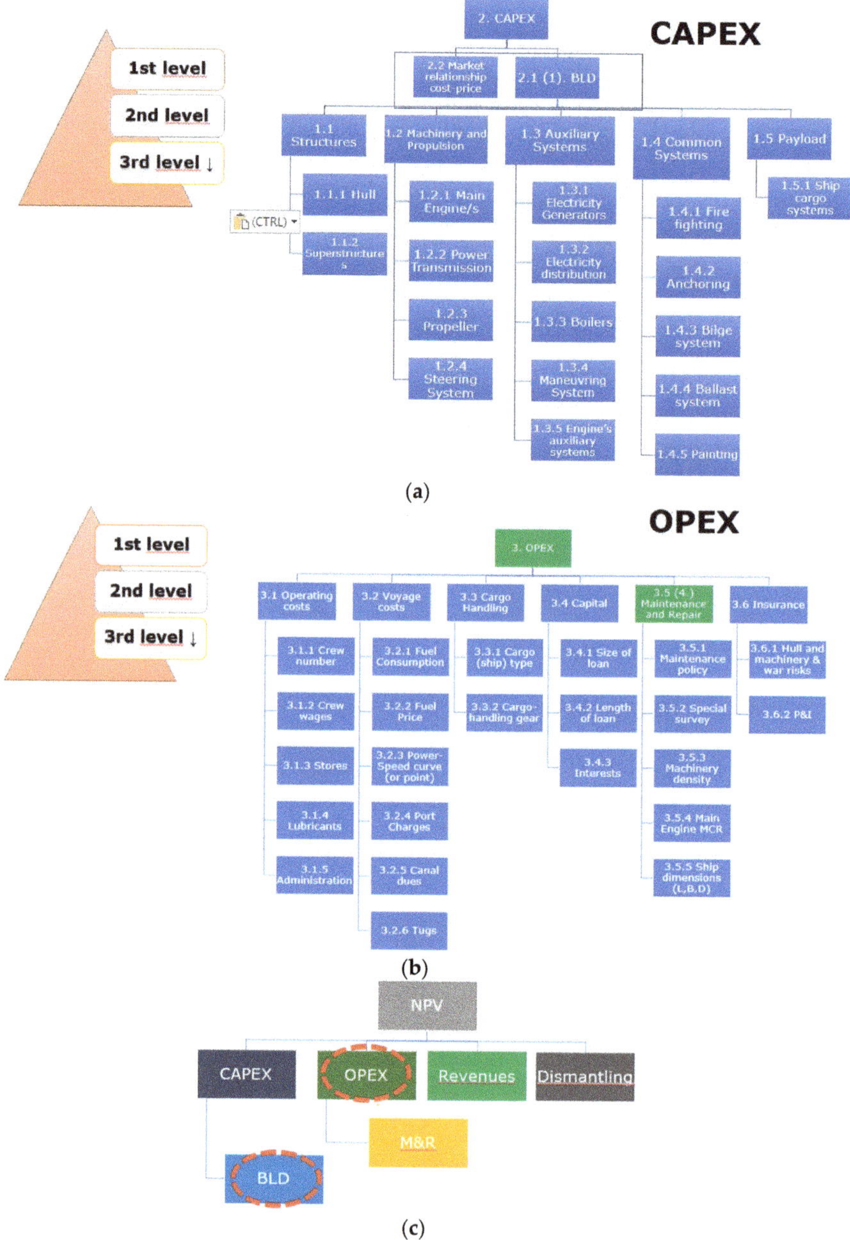

Figure 1. (**a**) CAPEX (Capital Expenditure) explanation; (**b**) OPEX (Operating Expenditure) explanation; P&I stands for Protection & Indemnity; MCR is the Maximum Continuous Rating; (**c**) NPV (Net Present Value) explanation. BLD: Building Cost; M&R: Maintenance and Repair costs.

A thorough explanation of the LCPA process is given in [7,10]. Nevertheless, in Figure 2, a block diagram of LCPA steps is provided. The same steps have been followed in the organization of the application case below.

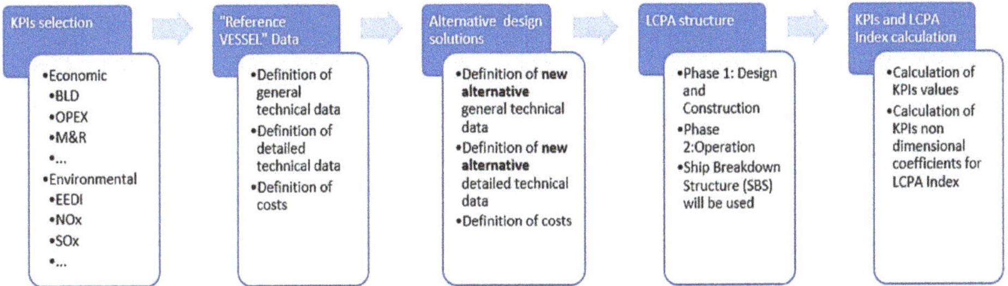

Figure 2. Block diagram illustrating LCPA (Life Cycle Performance Assessment) steps. KPIs: Key Performance Indicators; EEDI: Energy Efficiency Design Index.

In this work, the M&R costs evaluation have been improved to permit a better evaluation of the OPEX costs: the empiric formulation used in the current first version of LCPA tool could be replaced with a flexible prediction model based on real maintenance actions. The maintenance model, discussed in paragraph 4, would be implemented as part of the LCPA tool. A different energy system layout might also affect BLD value. In this work, such an issue was taken into account.

3. Maintenance Strategies

Maintenance actions ensure that a system performs in the best way during its whole life cycle, preserving its integrity and performances over time. Different maintenance techniques were developed in the last decades to better preserve system capabilities during its life cycle, minimizing the failure rate and downtime. All these actions can be summarized as follows [19].

Corrective maintenance: the simple one without any scheduled action. The operator attends when a failure occurs in switching off the system and performing the maintenance with more or less important economic implications. This approach is based on the belief that the costs sustained for downtime and repairs, in case of a fault, are lower than the investment required for a maintenance program. This strategy may be cost-effective until catastrophic faults occur.

Scheduled preventive maintenance: a step forward in maintenance policy. The manufacturer provides a so-called Mean Time Between Maintenance (MTBM) that is the best working time range when a maintenance action has to be performed to prevent system failure or degradation. In this way, an operator can plan the maintenance services to minimize the impact on working hours and so on costs and profits. The maintenance cycles are planned according to the need to take the device out of service. The incidence of operating failures is reduced. In a complex system with more sub-systems that work together to complete a task, this method can be a better way to plan the maintenance operations.

Performance-based maintenance: often indicated as Condition Based Maintenance (CBM), it is based on the response analysis of multiple sensors mounted on the system to measure actual working parameters like temperature, pressure, fluid levels, and more. Through prediction models, they are automatically compared with average values and performance indexes. Maintenance is carried out when some indicators give the signaling that the equipment is deteriorating and the failure likelihood is increasing. This strategy, in the long term, allows a drastic reduction in maintenance costs, thereby minimizing the occurrence of serious faults. With these previsions, the operator can plan one or more maintenance action only when it is requested and avoiding the interventions when it is not necessary.

These three actions are the most common, but other two maintenance policies can be further mentioned, namely the **adaptive maintenance** and the **perfective maintenance**. The first one is applied when the system needs to evolve and adapt for new needs or working contest to extend the system working life; the second one is performed when there is the chance to improve a system with the installation of innovative technology.

In this work, the scheduled preventive maintenance was assumed to develop the maintenance prediction model where the designer could evaluate the quality of the different configuration in relation to maintenance costs and time. This model would improve the current version implemented in the LCPA tool, based on an empirical formulation of the maintenance costs. Preventive maintenance is defined as a semi-deterministic model; however, it gives an effective way to compare different layouts and operational profiles at an early design stage: it takes into account, in fact, that designers could not have access to a large database, recorded dataset, or detailed information on systems they are investigating. The main information needed to set up a scheduled maintenance approach could be more easily obtained from manufacturer manuals. This is the reason why we used the preventive maintenance technique as a starting point [20,21]. The corrective maintenance and the dry-dock maintenance costs have been disregarded in this paper, and they are going to be taken into consideration later in the research activity.

4. Approach Description for Maintenance Costs Evaluation

As just described in the previous paragraph, the model for maintenance costs will be based on the scheduled preventive maintenance and the MTBM (Mean Time Between Maintenance) value [22]. Settled the ship type, the first step is to identify some different design configurations that satisfy the main owner requests, like speed, range, operational profiles, maneuverability performance, environment, and efficiency performance.

- The systems complexity and the huge number of sub-systems and single items installed in each assessed layout impose to choose a solid and structured procedure to evaluate the maintenance actions and related costs over the life cycle. From the builder point of view, the best practice would be to use the so-called Work Breakdown Structure (WBS), defined as a hierarchical and incremental decomposition of a project/system into smaller components: throughout a tree structure at each following iteration, a higher detail level is achieved, breaking up a complex system into a less complex one, and so on [23]. The operator decides when to stop dividing. This top-down structure allows to identify the elements by a single code number and to decide the ones useful in the analysis for the decided level of detail.
- For each selected system or subsystem, the maintenance tasks to include or exclude from the analysis are selected. The higher is the detail level to be considered, the more maintenance tasks have to be included in the tool.
- An MTBM value is needed for every single item selected for the analysis, expressed in working hours or years. The value is usually contained in the system manufacturer manuals with the corresponding maintenance task to be undertaken. The MTBM of an asset is the average length of operating time between one maintenance action and another, and it is usually based on a conservative stochastic distribution (Weibull distribution). Despite MTBM could be supposed a conservative value, if a system or item works out of its optimal working point (for a not negligible time), it is reasonable to suppose that the MTBM value might decrease.
- When defining alternative configurations, it is important to define the system working point. For example, considering a diesel propulsion engine, it is necessary to define its actual working condition expressed in term of MCR percentage and actual working hours. Starting from this information, the off-design working condition can be estimated together with is effects on the MTBM value delivered by the manufacturer. Combining off-design functioning hours and the corresponding power percentage enables to create a dimensionless corrective coefficient to update the actual maintenance plan and its impact on maintenance costs. This concept can be expressed through the proposed Formula (4) obtained through a try and error process based on the real data provided for the project:

$$\mathrm{MTBM'} = \mathrm{MTBM} - \frac{1}{\mathrm{h_T}} \cdot \left[\sum_i \mathrm{h}_{\mathrm{ACT}_i} \cdot \frac{\left(\mathrm{P}_{\mathrm{T}_i} - \mathrm{P}_{\mathrm{ACT}_i}\right)}{\mathrm{P}_{\mathrm{T}_i}} \right] \cdot \mathrm{MTBM} \tag{4}$$

where MTBM′ is the new corrected MTBM, the actual value; MTBM is the original manufacturer value; h_{ACT} is the number of actual off-design working hours; h_T is the number of total actual working hours in one year; P_T is the power corresponding to the optimal working point; P_{ACT} is the actual power; *i* is the operational scenario considered, like navigation at 13 kn or navigation at 8 kn, etc. This type of formulation can be applied to the diesel engines for the generation of propulsion.

If P_{ACT} and P_T are equal, the engine works at its best and MTBM = MTBM′. The same result is obtained if $h_{ACT} = 0$, i.e., the engine is in its best working point. We have assumed that the formulation has an application domain from 20% to 50% of MCR engines power: if an engine works above the 50% of its MCR, the corrective formula is not applied; at the same time for MCR less of 20%, the formula is not recommended.

- The next step requires to consider for each maintenance task the number of man-hours needed, the number of qualified technicians involved, and relevant costs (in €/h). All these values will be combined to calculate a total cost amount for every single task under analysis. If it is required, spare parts will be also included in the cost assessment for each task assumed in the tool.
- Once the main system's actual working hours and the ship life duration (in years) are inserted, the tool automatically calculates the number of maintenance action (times) during the whole life cycle (for each task) and the LCC for the maintenance activity defined as total costs per times. The costs refer to the real ship working conditions through MTBM′, and not to the ideal condition based on the manufacturer's MTBM. The MTBM or MTBM′ value is used to calculate how many times a particular maintenance action has to be performed during the ship life cycle. The "times" column in Figure 3 is obtained as (effective MTBM)/(item's total working hours in one year); the LCC for the maintenance activity is calculated as "times" column per "total costs" column in Figure 3.

The tool provides all the information needed to have a general overview of maintenance-related costs over the ship life cycle, with graphical and tabular results. Figure 2 summarizes the maintenance framework where WBS systems and subsystems are written in lines, and the higher is the level of detail, more lines are required.

The columns report (from left to right) a system or subsystem description; the maintenance task description (there could be more tasks for the same WBS voice); the number of same items; the effective MTBM, expressed in working hours or years and corrected with the previous formula; the number of hours dedicated to the task; the number of men required to do the particular task; the man costs; the spares cost (if a spare part is required); the total maintenance task costs; the times that the task is repeated during ship life cycle; the LCC value.

				Effective MTBM								Effective	
WBS	System Description	Task Description	n° items	Scheduled	Periodicity Limit	Man-hours	Man	Man Costs	Total	Spares Costs	Total Cost	Times	LCC
-	-	-	-	hours	years	hours	-	€/h	€	€	€		
200													
201 202 203 ...													
300													
301 302 303 ...													

Figure 3. Maintenance tool structure overview. LCC: Life Cycle Costing; MTBM: Mean Time Between Maintenance; WBS: Work Breakdown Structure.

The prediction model is structured to have the WBS, maintenance tasks, number of items, MTBM, man-hours, number of man, and man costs as input data. The model automatically calculates the

MTBM', the total cost of each maintenance tasks, the times each action has to be performed during the ship life, and the costs (indicated as LCC on the right in Figure 3).

During the early design stage, the definition of the ship operational profile is very relevant and has a strong influence on the propulsion system typology. In turn, the propulsion system strongly affects the ships final building cost and the costs linked to operational activity during the whole ship life [16]. From that described above, the LCPA tool has been further enhanced to evaluate the best configuration among some proposed alternative propulsion layouts. Due to the great number of systems and sub-systems installed, the complex connections between them, and the massive amount of required information, it is necessary to further reduce the domain of investigation in this development phase. This work aimed to develop a tool to predict the maintenance costs during the design stage, also when the information available is not so detailed. The intention was to develop a solid and adaptable starting point that could be improved during further interactions or directly customized from the designers themselves over their needs. Once the results of the costs derived from the maintenance tool are inserted in the main LCPA tool, the designer can provide a first attempt to quantify the total ship costs, not only related to building costs but also related to the operating costs. This type of assessment could permit to offer a better and complete overview of the product in a constructive discussion with the customer.

This methodology has been applied to a reference vessel, described in Section 5, to improve its propulsion system. The alternatives systems have been described one by one in Section 6 and are directly compared in Section 7 to identify the best solution.

5. Application Case: Reference Vessel Description

The methodology described in paragraph 4 is now applied on a reference vessel to evaluate propulsion system alternatives. The application platform is a research vessel designed and equipped to navigate the far reaches of the globe. These types of vessels support researches on the sea analyzing temperature gradients, sea chemical composition, carrying out biological or geophysics investigations, or bottom topography. The overall characteristics of a research vessel are identified by the scientific equipment, the specialized personnel on board, the required speed and range. The propulsion system has to be designed to ensure flexibility and a wide range of activities at different speeds. The original propulsion system of a model vessel and proposals of alternative configurations described in the following paragraph would be our starting points for investigations and analysis.

The main vessel dimensions and features are described in Table 1. The ship is equipped, in the original version, with two independent shaft lines, each provided with the main diesel engine, a small electric motor, a gearbox, and a controllable pitch propeller. Mechanical and electrical systems work together in the propulsion train, optimizing the ship propulsion efficiency and providing the right amount of power delivery to a propeller in any scenario.

Table 1. Main ship dimensions and performances.

Characteristic	Value	Units
Length over all	94	M
Length between perpendiculars	80	M
Beam	16.6	M
Depth	8.5	m
Displacement	3600	t
Draft	4.64	m
Cruise speed	13	Kn
Max speed	17	Kn
Range	3000	Nm

This hybrid propulsion system works as a CODELOD (COmbined Diesel and ELectric Or Diesel) where the only electric motors are used for low speeds (up 8 kts), and the diesel engines are instead used for high speeds (from 9 to 17 kts). The electric power generation at 400 V and 50 Hz, in all operating conditions, is ensured by three independent diesel generators connected to the main distribution grid. The emergency power generation is carried out by a dedicated diesel generator, located in a different small engine room. A simple functional scheme of the propulsion and electrical generation layout is shown in Figure 4.

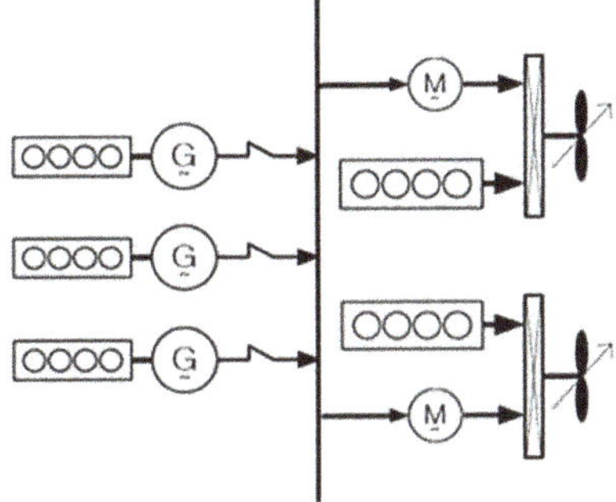

Figure 4. Original propulsion system layout.

The two propulsion engines are typical four strokes engines based on a supercharged diesel cycle with direct fuel injection. The supercharge is ensured by a turbo-compressor group driven by engine exhaust gases. The two diesel engines installed onboard can supply power output of 2289 kW, irreversible rotation, and opposite turns. A pneumatic system with compressed air is provided for the engines start, while the cooling system is made by a closed high-temperature freshwater circuit and a low temperature opened one.

The electric generation onboard is ensured by three gen-sets located in the main engine room. The electric power produced onboard is mainly used for the "payload" services (that include accommodation services but also scientific equipment power demand) and for the electric propulsion at low speed as well. The gen-sets can produce 650 kWe each at 1500 rpm with an electric output at 400 V and 60 Hz. A single gen-set is made up by a four strokes diesel engine with a direct fuel injection connected to an electric alternator. Table 2 provides a direct comparison between the propulsion types of diesel and the diesel gen-sets (MDO is the acronym of Marine Diesel Oil; MGO stands for Marine Gas Oil).

Table 2. Main Diesel.

Characteristic	Propulsion Diesel	Diesel Gen-Sets	Units
Number of cylinders	12 V type 45°	8 V type 90°	-
Maximum continuous rating	2289	650	kW
Rotation speed	1050	1500	g/min
Dimension	4100 × 1700 × 2760	4000 × 2100 × 200	mm
Weight	19000	10800	kg
Fuel supply	MDO/MGO	MDO	-

To increase energy efficiency in a range of low speed, the vessel is propelled by two electric synchronous type motor, one for each propulsion line. They can supply the maximum power output of 250 kW at 1500 rpm and an electric output at 400 V and 50 Hz. Main features are in Table 3.

It is important to define one or more profiles before starting the comparative analysis because they can strongly influence the results. As a research vessel, the ship will be optimized to work at two different speed range: low speeds during the maneuvering and oceanographic operations (typically from 0 to 8 knots) and high speeds during shift operation at design or maximum speed (from 12 to 17 knots).

Table 3. Electric motor.

Characteristic	Value	Units
Maximum power	250	kWe
Rotation speed	1500	g/min
Voltage output	400	V
Frequency output	50	Hz
Dimension	1160 × 670 × 860	mm
Weight	1030	kg

Figure 5 shows the operational profiles of the reference ship: the percentages refer to 165 days/year when the ship is operating at sea. In the remaining 150 days/year, the ship is considered in harbor, and in the remaining 50 days/year, in dry dock. In this scenario, 10% of working life (16 days) is spent at 0–5 kn; 35% (58 days) is spent at 6–8 kn in research operations; 5% (8 days in total) at 9–11 kn during speed transient phases; 35% (58 days) at 12–14 kn at design speed; 15% (25 days) at maximum speed.

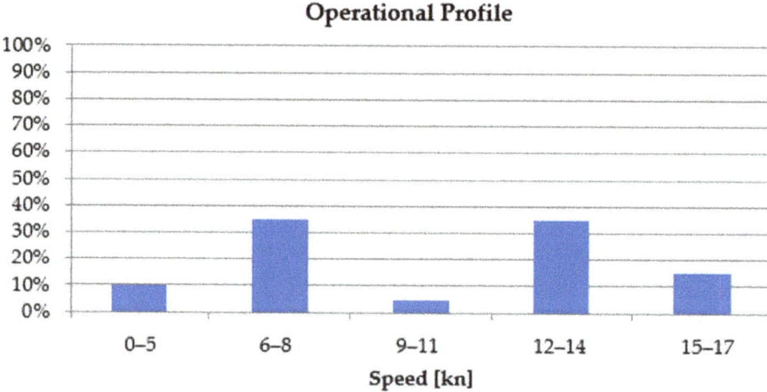

Figure 5. Reference ship operational profile.

All future considerations in this work would be based on this particular operating scenario.

To compare different design alternatives, the first step is to define how the vessel propulsion system works at a different speed and then to identify possible design alternatives to satisfy the operational profile most efficiently. These alternatives will be identified, during this first working phase, as composed by the elements listed below:

- Main Diesel Propulsion Engines (WBS 233)
- Electric Motors (WBS 235)
- Gearboxes (WBS 241)
- Shaft Lines and Bearings (WBS 243 and 244)
- Engines Sea-Water Cooling System (WBS 256)
- Fuel Supply System (WBS 261)
- Diesel Gen-Sets (WBS 311)
- Emergency Diesel Gen-Set (WBS 312)
- Heating, Ventilations and Air Conditioning HVAC System (WBS 514)
- Compressed Air System (WBS 551)

The flexibility of the tool structure ensures a further implementation of new WBS systems.

The operational profile has been related to a ship resistance prediction, to connect each speed to a propulsion power demand, and an electric balance, to better understand the total power demand during ship operations. The propulsion resistance prediction has been coupled to a controllable pitch propeller and the two propulsion diesel engines described above. A general overview of the ship propulsion system and ship power demand has been provided and used as a starting point.

Electric Load Balance

The term propulsion system (as we considered it in this work) is not only referred to the power required to move the ship, produced by two main diesel engines and two small electric motors, but it takes into account the electric production as well. This electrical power is needed not only for the electric propulsion motors, but also for the bow or stern thrusters, auxiliary propulsion systems, hotel loads, and the scientific equipment. In the original vessel configuration, this power demand is ensured by three diesel gen-sets (please refer to Figure 4 and Table 1) that satisfy the total electrical power demand in all ship operative conditions. The electric load balance of the reference vessel is available in Table 4. Utilization factors in Table 4 appear to be far from ideal figures, and, in particular, 95% and 88% relevant to summer harbor and winter harbor conditions are not in principle practicable, even though such values refer to a ship currently in operation. This configuration is, therefore, suitable for further improvement.

Table 4. Electric Balance of a research vessel.

Service	Summer Harbor	Winter Harbor	Summer Maneuver	Winter Maneuver	Summer Cruising	Winter Cruising
	0 kn	0 kn	1 to 8 kn	1 to 8 kn	9 to 17 kn	9 to 17 kn
	kW	**kW**	**kW**	**kW**	**kW**	**kW**
Propulsion system	12	12	91	91	59	59
Electric system	76	76	86	86	90	90
Control system	9	9	12	12	34	34
Aux system	341	292	374	308	390	348
Aux system	91	91	774	774	86	87
Outfitting	90	90	53	88	101	136
	619	**571**	**1390**	**1359**	**760**	**753**

	Summer Harbor	Winter Harbor	Summer Maneuver	Winter Maneuver	Summer Cruising	Winter Cruising
Gen-sets in use	1	1	3	3	2	2
Utilization factor	95%	88%	71%	70%	58%	58%

Harbor, Maneuver, and Navigation are the three phases considered in the electric balance. Each operational phase is split up in summer and winter condition to have an overview of power demand during different seasons.

For each column in Table 5, an indication of several gen-sets, in use, and their utilization factor has been provided. All this information gives an overview of how the propulsion configuration works in different scenarios during its working life. Besides, Table 5 shows for each scenario the number of working hours in a year of each WBS considered and described in paragraph 5.

Table 5. Working hours in a year for each WBS analyzed (see Abbreviation List at the end of the paper). Reference configuration.

Operation	Speed	n D/G	% D/G	WBS 233	WBS 235	WBS 241	WBS 243	WBS 256	WBS 261	WBS 311	WBS 514	WBS 551
	kn	-	-	hours	hours	hours	hours	hours	hours	hours	hours	hours
Port	0	1	0.95	0	0	0	0	0	1800	1260	2880	900
Manoeuvre	1 to 5	3	0.71	0	307.2	384	384	192	192	268.8	307.2	96
Research	6 to 8	3	0.70	0	1392	1392	1392	1113.6	696	1392	1113.6	348
Transient	9 to 11	2	0.58	96	0	192	192	153.6	192	134.4	153.6	48
Navigation	12 to 14	2	0.58	1392	0	1392	1392	1392	1392	974.4	1113.6	348
Navigation	15 to 17	2	0.58	600	0	600	600	600	600	420	480	150
				2088	1699.2	3960	3960	3451.2	4872	4449.6	6048	1890

Tables 4 and 5 refer to reference propulsion layout, and the information would be used as a solid base for the further considerations discussed in paragraph 6. All this conclude the preliminary phase.

6. Application Case: Design Alternatives

Starting from information identified in paragraph 5, it is possible to evaluate few alternative propulsion layouts to achieve a more efficient system, with possible advantages on building costs and operative costs, including all maintenance actions. In the following lines, three alternatives configurations are proposed: they have been designed focusing on a reduction in WBS working hours and/or maintenance costs.

6.1. First Design Alternative—Power Take-Off

The first alternative layout (identified as S1) proposes the introduction of a Power Take-Off (PTO) in the original propulsion system. In particular, the PTO is supplied by the main propulsion Diesel with the double aim to reduce the working hours of one or more diesel gen-sets, and to achieve a better working point both for diesel engines and for the gen-sets (compared to the original layout). Engines power size are the same as reference vessel, and the PTO is active only from 9 kn to 17 kn; up to 8 kn, the configuration works exactly as the original one. So, for the harbor, maneuvering, and navigation activities up to 8 kn, the power demand is supplied by two electric motors (for propulsion and maneuvering) and by the diesel gen-sets, as reported in Table 6. After 8 kn, the propulsion starts to be supported by the diesel engines, so, after this point, it is possible to use the electric motors (used as PTO) instead of the diesel gen-sets to supply the power needed for hotel services. Figure 6 clarifies this working operation.

Table 6. First alternative configuration power-use analysis results. Focus on changed WBS working hours.

Speed	P_B	PTO	n D/G	% D/G	WBS 235	WBS 311
kn	kW	kW	-	-	hours	hours
0	0	0	1	0.95	0	1260
1 to 5	83	0	3	0.71	307.2	268.8
6 to 8	320	0	3	0.70	1392	1392
9 to 11	807	500	1	0.41	192	67.2
12 to 14	1760	500	1	0.41	1392	487.2
15 to 17	3617	500	1	0.41	480	240
					3763.2	3715.2

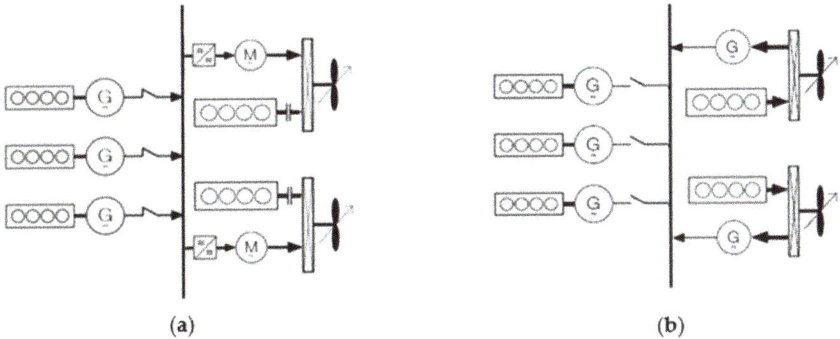

Figure 6. Power generation layout in the first alternative configuration proposed. (**a**) How configuration works up to 8 kn: the two electric motors supply the total propulsion power, while the gen-sets supplies the hotel power. The Diesel engines are disconnected. (**b**) How configuration works from 9 kn to 17 kn: the two diesel engines supply both propulsion power and the PTO (Power Take-Off) units. The PTO delivers electrical power to the main switchboard, while the surplus in the power demand is covered by one or more gen-sets.

Table 6 reports the configuration analysis in terms of propulsion power (P_B), PTO, number of active gen-sets and their working point (%), and WBS voices that have changed their working hours. The analysis put in evidence the PTO power impact during navigation phase: after 8 kn, the PTO supplies more electrical power in the network, so it is possible to use only one gen-set, compared to two in the original configuration. There in an increase in for WBS 235 working hours, while a decrease is evident for WBS 311 working hours.

It is important to underline that the gen-set in use works at 41.5% of its nominal power, and for diesel gen-sets, it is not a good solution in terms of fuel consumption and maintenance. This consideration took us to consider a second alternative configuration to solve this problem.

6.2. Second Design Alternative—Power Take-Off with Higher Power Size

The second alternative (S2) layout proposes the introduction of a higher power size PTO and also a higher power size for gen-sets to obtain an efficient working point. This solution could be identified as an evolution of the previous one, with the main aim to reduce to zero the number of gen-sets used during the navigation phase. So, the configuration layout is the same, as defined in 6.1, and differs only for the items size and working point. Figure 6 continues to represent the system layout. The new proposal increases the electric motors size (also used as PTO) from 250 kW to 390 kW and gen-sets size from 650 kW to 850 kW. This option permits to satisfy the total amount of power demand in the navigation phase using the PTO only. The total number of simultaneous active gen-sets, in the worse scenario, decreases from three to two, but from a reliability and redundancy perspective, the third gen-sets are used as a stand-by element. The results achieved in this way have been represented in Table 7. It puts in evidence how the PTO supplies the total electric power demand during the navigation phase over 8 kn. The two diesel engines are the only active power generators during the navigation from 9 to 16 kn, both in summer and winter.

Table 7. Second alternative configuration power-use analysis results. Focus on changed WBS working hours.

Speed	P_B	PTO	n D/G	% D/G	WBS 235	WBS 311
kn	kW	kW	-	-	hours	hours
0	0	0	1	0.850	0	1260
1 to 5	83	0	2	0.850	307.2	268.8
6 to 8	320	0	2	0.810	1392	974.4
9 to 11	807	780	0	-	192	0
12 to 14	1760	780	0	-	1392	0
15 to 17	3617	780	0	0.600	480	120
					3763.2	2623.2

Comparing Table 7 to Table 5 (S0 configuration), the higher power sizes ensured a better gen-sets working point in all conditions, less working hours, a reduction in fuel consumption, and stress on the mechanical elements. A reduction of working items was ensured: from three to two in maneuver; from two to 0 during navigation. The gen-sets total working hours were also reduced (from 4449 to 2623), but the working hours of electric motors were increased from 1699 to 3763. Main diesel working hours were the same, as described in Tables 5 and 6, but the working point was higher due to the PTO supply during navigation from 8 kn to 17 kn.

The remaining systems have the same working hours as S0, but the active units are lesser with a positive influence on maintenance tasks. All this information are the input data to the maintenance tool described in paragraph 4. The comparison in terms of a maintenance plan, between reference layout S0 and layout S2, is shown in Figure 7. The blue line is referred to reference layout (S0), while the red one to second alternative layout (S2): over 20 years, the S2 maintenance actions are, on average, less expensive than the original layout. The MTBM correction affects not only the distribution of the costs over the years but also the maintenance periodicity, that is out of phase than the original maintenance plan. From the owner point of view, this type of results presentation provides an overview of M&R, permitting to schedule when could be better to sell the ship to prevent excessive M&R costs.

Paragraph 7 provides a costs comparison between all alternative solutions described here.

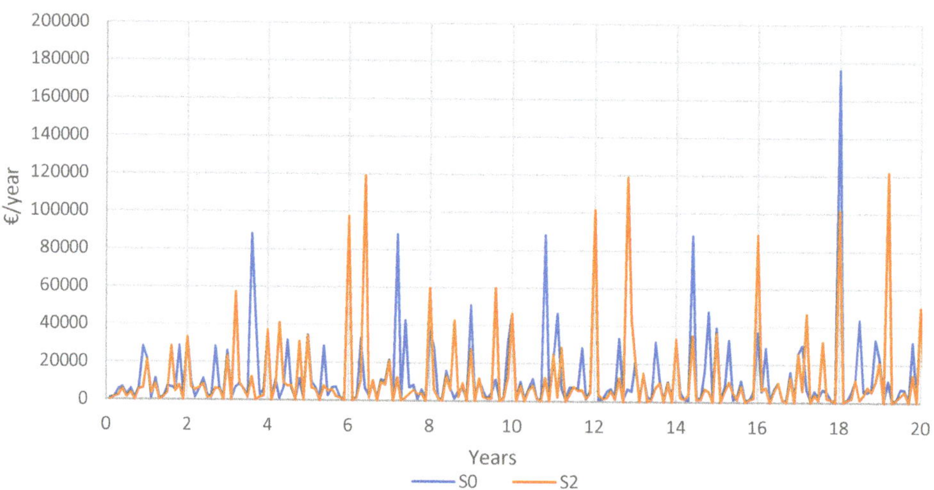

Figure 7. Second layout M&R costs.

6.3. Third Design Alternative—Total Electric

The third alternative layout (S3) proposes a complete change in the propulsion layouts through the introduction of fully electric propulsion. Instead of the two main Diesel engines, two main electric motors have been installed, supplied by four diesel gen-sets. This configuration could be an optimal choice for ships that have a special environment request, as a research vessel, and need a large amount of electric power available during the operational phases. A comparison between hybrid propulsion and fully electric propulsion, on maintenance perspective, could be interesting. As synthesized in Figure 8, usually this configuration is composed of diesel gen-sets; main switchboards; propulsion transformers and frequency converters; electric engines for propulsion; gearboxes (optional) and propellers.

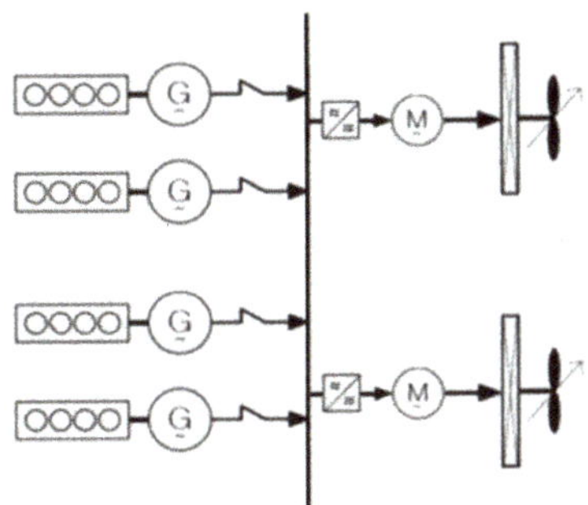

Figure 8. Full electric propulsion system layout.

This configuration ensures high power flexibility, and so it is not necessary to divide the speed range as done before. From 1 to 17 kn, the total electrical power demand will be satisfied by a variable number of gen-sets, regulated by onboard automation. The electrical motors are only used as main propulsion engines: a PTO is not considered in this configuration.

The two electric propulsion engines have a power size of 1800 kW each, to ensure a 17 kn maximum speed. The gen-sets have to satisfy the power requested at 17 speed (by propulsion) and the power requested by the hotel loads during navigation, equal to 760 kW in summer navigation (worse condition). So, the gen-sets are four units of 1150 kW each. This layout ensures the redundancy of power generation units and, at the same time, permits a high control over the electrical power generation during the different operational phases. Table 8 proposes an overview of WBS that changed their working hours.

Table 8. Third alternative configuration power-use analysis results. Focus on changed WBS working hours.

Speed	P_B	PTO	n D/G	% D/G	WBS 233	WBS 235	WBS 311
kn	kW	kW	-	-	hours	hours	hours
0	0	0	1	0.550	0	0	900
1 to 5	83	0	2	0.650	0	384	192
6 to 8	320	0	2	0.870	0	1392	417.6
9 to 11	807	0	2	0.700	0	192	96
12 to 14	1760	0	3	0.680	0	1392	1044
15 to 17	3617	0	4	0.960	0	600	600
					0	3960	3249.6

Figure 9 provides an overview of the comparison of a maintenance plan between S0 and S3. M&R costs are, on average, less expensive than the original ones with a better peaks distribution over the years. Plus, a fully electric solution is eco-friendly, that is not a negligible aspect for a research vessel.

The costs comparison is proposed in paragraph 7.

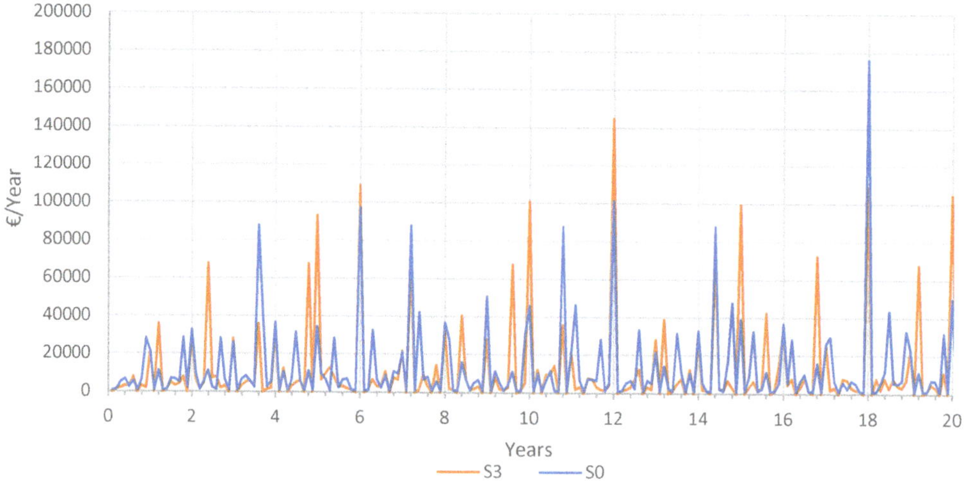

Figure 9. Third layout M&R costs.

7. Application Case: Results

Four different solutions were analyzed in this paper: the traditional diesel configuration (S0), the PTO solution (S1), the PTO of higher power size, and the fully electric one (S3). Table 9 provides the results of the total costs obtained with a complete LCPA analysis: we mainly focused on BLD, OPEX, and M&R values to compare the solutions. EEDI, Sox, and NOx are approximated values, but they can give an idea of the solution quality. Speaking in terms of BLD, the reference ship layout (S0) is the less expensive, while the fully electric configuration, as expected, is the most expensive; solution S1 and S2 are very similar, S2 is penalized by the third gen-sets installed for a redundancy aspect but not needed in terms of power generation (two gen-sets covered all scenarios). In terms of OPEX, the best solution is S2, while the reference configuration (S0) is the worse. It is important to remind that OPEX value includes fuel consumption, so, this value could also be read as, S0 has a higher fuel consumption than S2 and S3. For each configuration, fuel consumption has been calculated, combining the utilization factors and the complete engines diagrams. Referring to data reported in Tables 5–8, the results obtained are truthful.

The M&R are lower in S2 and higher in S1: as expected, S2 is the configuration that has been designed to ensure engines' better working point and this is underlined in OPEX and M&R values that are the best. S1 that has a worse working point, as shown in Table 8, also has higher M&R costs: a bad engine utilization reduces the MTBM and increases the maintenance times over ship life cycle.

Table 9. Results of the layout comparison.

Data		Reference S0	S1	S2	S3
BLD	[€]	70,038,000	70,697,980	72,715,388	76,714,838
CAPEX	[€]	98,877,176	99,808,913	102,657,018	108,303,301
OPEX	[€/year]	4,026,187	3,937,086	3,820,971	3,980,591
M&R	[€/year]	452,777	476,177	365,856	409,999
NPV	[€]	205,493,593	205,369,584	205,584,835	214,916,381
EEDI	[gCO2/t*kn]	266.63	251.09	260.68	260.68
SOx	[gSOx/t*kn]	4.99	4.70	4.88	4.88
NOx	[gNOx/t*kn]	11.64	10.96	11.38	11.38

The EEDI value points out that all alternative configurations have better efficiency and lower emissions than the reference one. S1 is the solution with the lower EEDI value, but we should underline that value in S2 is affected by the third gen-set installed as a stand-by unit. The EEDI value is a function of installed power.

It is important to note that despite S1 has worse engines working point, and therefore higher M&R is the best solution. Speaking in terms of NPV, the total value not differs so much from the other ones.

In Figure 10, it has been reported the LCC values maintenance over time for the four layouts S0, S1, S2, S3. As just said, solution S2 is the best in terms of operational maintenance; indeed, it has been designed to minimize maintenance costs. An unexpected result is given by solution S0: it is a cheaper solution than S1 even though the main engines are used less efficiently than in solution S1 (characterized by the introduction of the PTO).

From these results, it is also possible to see that the fully electric, besides being the most expensive solution, will not guarantee a significant advantage during the years. From this analysis, we can also say that, at least for our application case, it is not true that the most expensive solution is always the best in terms of maintenance costs. Analyzing the last years of the ship life in the graph (25–30 years), it seems that solution S2 would bring a good saving compared to the original solution.

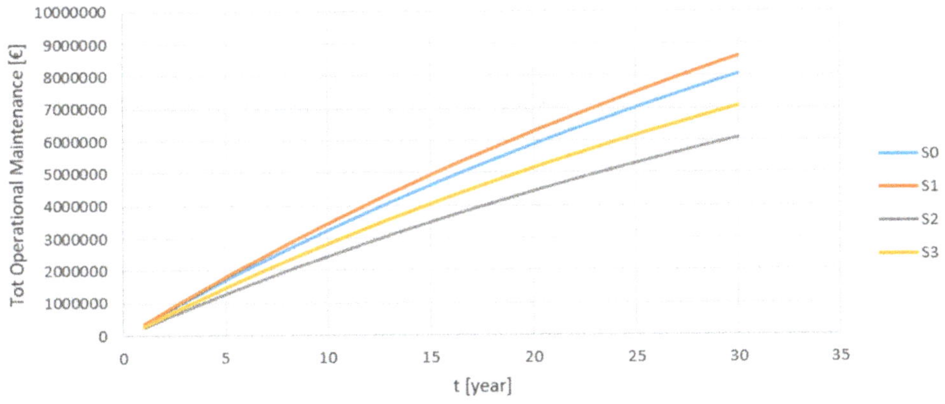

Figure 10. Integration over 30 years of Operational Maintenance costs.

Table 10 provides the results in terms of dimensionless index ready for the LCPA calculations: the coefficient is always defined between zero (worst) and one (best). S1 is the best, and S2 is the second-best solution, even if it is affected by the power surplus installed. Not purchasing the third gen-set, decision subjected to owner agreement, implies less BLD and M&R. The S3 is the worst layout, but it is important to stress that LCPA tool does not take into account noise and vibration problems, new possible weights distribution onboard, engine room locations, and off-limits routes.

The last column in Table 10 is part of the LCPA tool structure: for each KPI, a weight has been assigned. The assignation is arbitrary and given by the designer or the owner or the shipbuilder. In this case, we selected BLD and M&R as most interesting KPIs, followed by OPEX. In this perspective, despite NPV value is one of the most popular economic indicators, it has not been deeply discussed and analyzed during this work focused on maintenance actions and costs: the weight assigned in Table 10 is equal to 0 for this reason. The NPV value does not present significant changes in S0, S1, and S2 because the technology used in the propulsion system is the same (i.e., diesel engines), while the S3 configuration employs large electric motors as main propulsion item.

Table 10. Results of the layout comparison.

Data	Reference S0	S1	S2	S3	Weight
BLD	1.00	0.90	0.60	0	0.4
CAPEX	1.00	0.90	0.60	0	0
OPEX	0.00	0.43	1.00	0.22	0.2
M&R	0.21	0	1.00	0.60	0.4
NPV	0.99	1.00	0.98	0.00	0
EEDI	0	1.00	0.38	0.38	0.25
SOx	0	1.00	0.38	0.38	0.5
NOx	0	1.00	0.38	0.38	0.25

Index	Reference S0	S1	S2	S3	Weight
LCC Index	0.48	0.45	0.84	0.28	0.5
LCA Index	0.00	1.00	0.38	0.38	0.5
LCPA Index	0.24	0.72	0.61	0.33	-

The same results are shown in Figure 11 below.

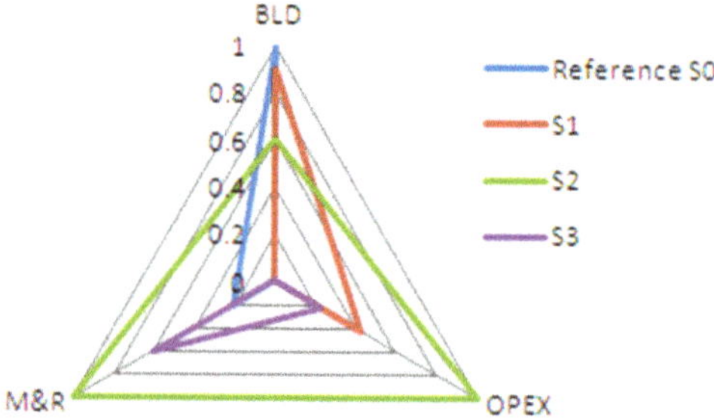

Figure 11. KPIs coefficients in spider graph.

8. Conclusions

For a research vessel, all the necessary data have been provided to start developing and assess a maintenance prediction model based on a real maintenance plan and not on a parametric formulation. The methodology described in Section 4 is a result of multiple interactions: it has been developed for the reference vessel only and then readapted to a general application for various vessel types. Figure 3 provides a simplified scheme of the maintenance tool and calculation structure.

During the tool development, a corrective Formula (4) for diesel engines MTBM has been defined to take into account design mistakeS on the maintenance costs. Application cases have been carried out on a research vessel, as exposed in Sections 5 and 6. Different propulsion solutions have been identified, and the tool can characterize them in terms of different performances; but, at the same time, it not so simple to identify the best solution because of the final evaluation that is strictly related to the stakeholder point of view.

Further improvements could derive from the following activity:

- to improve the modeling of corrective maintenance actions and dry dock actions;
- to improve Formula (4) with a major number of real maintenance data and with the implementation of engines overfeed curves.

Author Contributions: For conceptualization, methodology, validation, formal analysis and investigation P.G., G.F., M.M. and G.M. gave their contribution. For software and data curation, G.F., M.M. and G.M. gave their contribution. The writing—original draft preparation was made by G.M. and the writing—review and editing has been carried out by P.G. and G.F. Resources have been provided by M.M. Supervision has been carried out by P.G., G.F. and M.M. Project administration and funding acquisition was in charge of P.G.

Funding: The reported research is still under development in the frame of the HOLISHIP (HOLIstic optimization of SHIP design and operation for life cycle) project (2016–2020), which has been funded by the European Commission in the HORIZON 2020 Transport Program, Grant Agreement nr 689074.

Acknowledgments: The Authors would like to express their gratitude to Alberto Dachà (Fincantieri s.p.a) for the valuable contribution he gave in terms of constructive discussion and advise during the development of the research activity.

Conflicts of Interest: The authors declare no conflict of interest and the funders had no role in the design of the study; in the collection, analyses, or interpretation of data; in the writing of the manuscript, or in the decision to publish the results.

Abbreviations List

BLD	Building Costs
CAPEX	Capital Expenditure
CODELOD	COmbine Diesel and ELectric Or Diesel
EEDI	Energy Efficiency Design Index
KPI	Key Performance Index
LCA	Life Cycle Assessment
LCC	Life Cycle Costing
LCPA	Life Cycle Performance Assessment
MTBM	Mean Time Between Maintenance
M&R	Maintenance and Repair
NPV	Net Present Value
OPEX	Operating Expenditure
PTO	Power Take-Off
S0	Solution 0 or Reference Configuration
S1	Solution 1
S2	Solution 2
S3	Solution 3
WBS 233	Main Diesel Propulsion Engines
WBS 235	Electric Motors
WBS 241	Gearboxes
WBS 243	Shaft Lines
WBS 244	Bearings
WBS 256	Engines Sea-Water Cooling System
WBS 261	Fuel Supply System
WBS 311	Diesel Gen-Sets
WBS 312	Emergency Diesel Gen-Set
WBS 514	HVAC System
WBS 551	Compressed Air System

References

1. Jeong, B.; Jang, H.; Zhou, P.; Lee, J. Investigation on the marine LNG propulsion system for LNG carriers through an enhanced hybrid decision-making model. *J. Clean. Prod.* **2019**, *230*, 98–115. [CrossRef]
2. Jeong, B.; Wang, H.; Oguz, E.; Zhou, P. An effective framework for life and cost assessment for marine vessels aiming to select optimal propulsion systems. *J. Clean. Prod.* **2018**, *187*, 111–130. [CrossRef]
3. Trivyza, N.L.; Rentizelas, A.; Theotokatos, G. A novel multi-objective decision support method for ship energy systems synthesis to enhance sustainability. *Energy Convers. Manag.* **2018**, *168*, 128–149. [CrossRef]
4. Dimopoulos, G.; Georgopoulou, C.; Stefanatos, I.; Zymaris, A.; Kakalis, N. A general-purpose process modelling framework for marine energy systems. *Energy Convers. Manag.* **2014**, *86*, 325–339. [CrossRef]
5. Livanos, G.; Theotokatos, G.; Pagonis, D.N. Techno-economic investigation of alternative propulsion plants for Ferries and RoRo ships. *Energy Convers. Manag.* **2014**, *79*, 640–651. [CrossRef]

6. Papanikolaou, A. *A Holistic Approach to Ship Design, Volume 1: Optimisation of Ship Design and Operation for Life Cycle, Chapter 1*; Springer: Cham, Switzerland, 2019.
7. Gualeni, P.; Maggioncalda, M. Life cycle ship performance assessment (LCPA): A blended formulation between costs and environmental aspects for early design stage. *Int. Shipbuild. Prog.* **2018**, *65*, 127–147. [CrossRef]
8. Papanikolaou, A. *A Holistic Approach to Ship Design, Volume 1: Optimisation of Ship Design and Operation for Life Cycle, Chapter 12*; Springer: Cham, Switzerland, 2019.
9. Langdon, D. *Literature Review of Life Cycle Costing (LCC) and Life Cycle Assessment (LCA)*; Davis Langdon Management Consulting: London, UK, 2016.
10. Maggioncalda, M.; Gualeni, P.; Notaro, C.; Cau, C.; Bonazountas, M.; Stamatis, S. Life Cycle Performance Assessment (LCPA) Tools. In *A Holistic Approach to Ship Design*; Papanikolaou, A., Ed.; Springer: Cham, Switzerland, 2019; pp. 383–408.
11. Bolbot, V.; Trivyza, N.L.; Theotokatos, G.; Boulougouris, E.; Rentizelas, A.; Vassalos, G. Cruise ship optimal power plants design identification and quantitative safety assessment. In *2nd International Conference on Modelling and Optimisation of Ship Energy Systems*; Theotokatos, G., Corradu, A., Eds.; University of Strathclyde: Glasgow, UK, 2019; pp. 55–64.
12. Jeong, B.; Oguz, E.; Wang, H.; Zhou, P. Multi-criteria decision-making for marine propulsion: Hybrid, diesel electric and diesel mechanical systems from cost-environment-risk perspectives. *Appl. Energy* **2018**, *230*, 1065–1081. [CrossRef]
13. Shetelig, H. Shipbuilding Cost Estimation. Master's Thesis, University of Trondheim, Trondheim, Norway, 2013.
14. Stopford, M. *Maritime Economics*, 3rd ed.; Routledge: Abingdon on Thames, UK, 2003.
15. Lamb, T. *Ship Design and Construction*, 3rd ed.; Society of Naval Architects and Marine Engineers: Alexandria, VA, USA, 2003; Volume 1.
16. International Maritime Organization. *IMO (International Maritime Organization) Resolution MEPC.212 (63)-Guidelines on the Method of Calculation of the Attained Energy Efficiency Index (EEDI) for New Ships*; International Maritime Organization: London, UK, 2012.
17. International Maritime Organization. *IMO (International Maritime Organization) Resolution MEPC.176 (58), MARPOL Annex VI-Prevention of Air Pollution from Ships*; International Maritime Organization: London, UK, 2008.
18. Straub, A. Maintenance and Repair. In *International Encyclopedia of Housing and Home*; Elsevier: Oxford, UK, 2012; Volume 4, pp. 186–194.
19. Turan, O.; Ölçer, A.I.; Lazakis, I.; Rigo, P.; Caprace, J.D. Maintenance/repair and production-oriented life cycle cost/earning model for ship structural optimisation during conceptual design stage. *Ships Offshore Struct.* **2009**, *4*, 107–112. [CrossRef]
20. Butman, B. *Fundamentals of Ship Maintenance and Repair for Future Marine Engineers*; Merchant Marine Academy, Kings Point: New York, NY, USA, Jan 2014.
21. Knight, A. *MTBF, MTTR & MTBM, Reliability Metrics*; IRISS: Bradenton, FL, USA, 2017.
22. Norman, E.S.; Brotherton, S.A.; Fried, R.T. *Work Breakdown Structures: The Foundation for Project Management Excellence*; John Wiley & Sons: Hoboken, NJ, USA, 2008.
23. Stouffer, V. Use of Weibull Failure Rates. In Proceedings of the SCEA/ISPA Workshop, St. Louis, MO, USA, 2–5 June 2009.

Article

Life Cycle Assessment of LNG Fueled Vessel in Domestic Services

Sangsoo Hwang [1,2], Byongug Jeong [1,*], Kwanghyo Jung [2], Mingyu Kim [3] and Peilin Zhou [1,4]

1 Department of Naval Architecture, Ocean and Marine Engineering, University of Strathclyde, Glasgow G4 0LZ, UK; ssangsoo82@naver.com (S.H.); peilin.zhou@strath.ac.uk (P.Z.)
2 Department of Naval Architecture and Ocean Engineering, Pusan National University, Busan 46241, Korea; kjung@pusan.ac.kr
3 Maritime Research Institute, Korea Center for International Maritime Safety Cooperation, Sejong 30103, Korea; mingyu.kim@imkmc.or.kr
4 Faculty of Marine Science and Technology, Harbin Institute of Technology, Weihai 264209, China
* Correspondence: byongug.jeong@strath.ac.uk; Tel.: +44-074-256-94809

Received: 27 August 2019; Accepted: 5 October 2019; Published: 10 October 2019

Abstract: This research was focused on a comparative analysis of using LNG as a marine fuel with a conventional marine gas oil (MGO) from an environmental point of view. A case study was performed using a 50K bulk carrier engaged in domestic services in South Korea. Considering the energy exporting market for South Korea, the fuel supply chain was designed with the two largest suppliers: Middle East (LNG-Qatar/MGO-Saudi Arabia) and U.S. The life cycle of each fuel type was categorized into three stages: Well-to-Tank (WtT), Tank-to-Wake (TtW), and Well-to-Wake (WtW). With the process modelling, the environmental impact of each stage was analyzed based on the five environmental impact categorizes: Global Warming Potential (GWP), Acidification Potential (AP), Photochemical Potential (POCP), Eutrophication Potential (EP) and Particulate Matter (PM). Analysis results reveal that emission levels for the LNG cases are significantly lower than the MGO cases in all potential impact categories. Particularly, Case 1 (LNG import to Korea from Qatar) is identified as the best option as producing the lowest emission levels per 1.0×10^7 MJ of fuel consumption: 977 tonnages of CO_2 equivalent (for GWP), 1.76 tonnages of SO2 equivalent (for AP), 1.18 tonnages of N equivalent (for EP), 4.28 tonnages of NMVOC equivalent (for POCP) and 26 kg of PM 2.5 equivalent (for PM). On the other hand, the results also point out that the selection of the fuel supply routes could be an important factor contributing to emission levels since longer distances for freight transportation result in more emissions. It is worth noting that the life cycle assessment can offer us better understanding of holistic emission levels contributed by marine fuels from the cradle to the grave, which are highly believed to remedy the shortcomings of current marine emission indicators.

Keywords: LNG-fueled ship; IMO GHG; LNG; MGO; LCA; marine fuel

---

1. Introduction

Today, humanity is in the age of the most prosperous history. The advancement of industrial technology has allowed people to share goods with no barriers. Seaborne trade significantly contributes to this trend. In 2017, about 10.7 billion tons of products were traded through water, which represents an enormous amount of energy consumption, thereby producing emissions recklessly [1].

Given this, the International Maritime Organization (IMO) has developed a set of stringent regulations on emission control. In particular, MARPOL Annex VI Reg. 14 introduces a progressive reduction in sulfur content contained in marine fuels by 1 January 2020 when the sulphur content in those fuels should be reduced to 0.5% m/m (mass/mass) in the non-ECAs (emission control areas) as illustrated in Figure 1. For the ECAs, the emission levels have been curbed to 0.1% m/m since 2015.

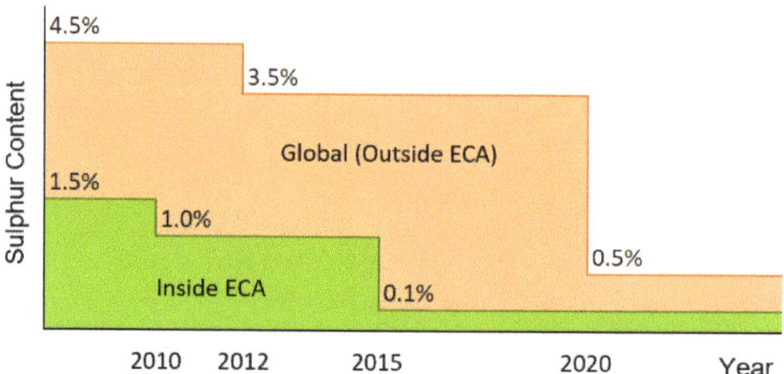

Figure 1. IMO Sulphur Regulation (MARPOL Annex VI Reg. 14).

Since conventional marine petroleum products cannot meet these regulations, marine engineers and shipowners are turning their attention to alternative fuel sources. Currently, liquefied natural gas (LNG) is considered as one of the most credible alternative marine fuels that can meet the upcoming air pollution regulations with respect to minimizing sulfur oxides (SOx), nitrogen oxides (NOx), and particulate matter (PM). As a result, LNG-fueled vessels have been gradually introduced in the marine industry and their number now reaches to over 100 ships [2].

Various studies have shown the environmental benefits of using LNG in ship operations [2–6]. However, the mainstream of those past studies largely focused on onboard emission levels at sea, whereas a considerable amount of emissions ranging from extraction to the transportation to the end user was ignored. To acquire better understanding of the impact of LNG on the emission levels as a whole, there still needs of developing systematic and comprehensive approach to evaluate the environmental impact of LNG fuel from a lifecycle perspective.

Meanwhile, the current marine environmental calculators, known as Energy Efficiency Design Indicator (EEDI) and Energy Efficiency Operating Indicator (EEOI), are considered not practicable in terms of understanding the holistic environmental impacts of marine fuels. It is because those indicators are only focused on calculating the emissions produced from onboard fuel combustion. In other words, those indicators are technically ignorant emissions generated from other life cycle processes of marine fuels.

This analytic limitation not only leads to inaccuracies in the calculation, but also mis-guide us to wrong conclusions when choosing clean marine fuel. For instance, they may indicate the hydrogen applied to marine fuel cells as the cleanest fuel source simply because it produces the least level of emission during ship operation. Interestingly, the hydrogen may be generated from LNG which is regarded more harmful fuel source over the hydrogen. If extending our view from the final use to the fuel production and supply chains, the results cannot be answered as simple as the conventional indicators tell us.

To remedy this issue, the International Maritime Organization (IMO) and member states have been determined to develop new guidelines for estimating the life cycle environmental impacts of marine fuels through the document of IMO MEPC.308 (73) [7]. To respond to this resolution, the initial idea was proposed with the document of IMO ISWG GHG, 5/4/5 [8], which requires far extensive follow-up research and case studies as future works.

Given this background, this paper was motivated to introduce an approach of LCA and to demonstrate its effectiveness through a comparative LCA of LNG with MGO in practical supply chain cases.

2. Literature Review

In effort to investigate the holistic environmental impacts of shipping-related issues, the concept of the life cycle assessment (LCA) has been applied to various studies over decades. There are some remarkable researches worth being mentioned in a methodological point of view. Guinée [9] presented a handbook for guiding to apply International Organization for Standardization (ISO) for LCA analysis and Finnveden et al. [10] discussed the recent development and trends of the LCA applied to industrial studies. Dynamic LCA in consideration of time domain was introduced by Levasseur et al. [11]. These literature provide high-quality basement for life cycle assessment. Nevertheless, Woods et al. [12] pointed out the lack of the LCA studies on investigating the marine environment impacts, addressing the scarcity of LCA modelling to quantify the effects of products and processes on marine biodiversity.

Not surprisingly, there are voluminous LCA studies evaluating the environmental impacts of LNG. Some representative examples are noteworthy. Bengtsson et al. [13] applied the LCA for a comparative study across crude oil, LNG and other competitive marine fuels. The research results revealed that LNG would have a relatively lower GWP compared to other candidates. Thinkstep [14] also provided comparative life cycle assessment for the natural gas with other marine fuels in terms of GHG intensity. Similar study was carried out by El-Houjeiri et al. [15] which compared Saudi Crude oil to the natural gas in other regions in terms of GHG emission. Also, Sharafian et al. [16] provided research about GHG and other air pollutants for natural gas. A localized GHG emission taking into account upstream life cycle was studied by Liu [17] where climate change impact by supply of natural gas in Western Canada was studied. This study provides a suggestion to find the prospect that localized impact for the same fuel can vary in different industries. However, those research were largely focused on the GHG impact and lacked discussion of other impact potentials of local pollutants such as AP, EP, POCP and PM. In addition, the diversity of supply routes was not included in the research scope.

Some interesting LCA studies were also conducted to evaluate local pollutants associated with LNG. A study by Brynolf et al. [18] compared the environmental impact of LNG to future marine fuels: liquefied biogas, methanol and biomethanol. An improved method to reduce net climate change was suggested in the research. Life cycle inventory and analysis of fuels in Singapore was presented in Tan et al. [19] that discussed several types of emissions from Singapore power plants compared between LNG and diesel oil. Tagliaferri et al. [20] investigated LNG transport from Qatar to UK with detailed and diverse scenarios. Jeong [21] provided the holistic research for HFO, MGO and LNG in terms of GWP, AP, EP, POCP.

In particular, it is noteworthy that LCA research has been extended not only to fuel types, but also to the ship building field. Hua et al. [22] analyzed the total life cycle emissions of a post-Panamax container ship running in both HFO and natural gas. Jeong et al. [23] presented the excellence of using LNG-fueled engines based on economic and environmental viewpoints. In the research, the life cycle cost assessment (LCCA) was advised as a useful tool for decision making across industries. It offered the possibility of extending the LCA in the economic point of view, since cost impacts cannot be a negligible issue for the marine industry. Rocco et al. [24] studied the purification process of LNG in the LCA point of view. Miksch [25] analyzed the shipping routes from United States (U.S.) to Asia transporting LNG. This research showed that the LNG supply chain would be sensitive to economic and environmental impacts of fuels. In addition, Dong and Cai [26] discussed several ways to reduce the environmental impact by reducing the fuel consumption rate.

Studies on future marine fuels integrated with advanced technologies rather than LNG cannot be neglected. Alkaner and Zhou [27] presented comparative life cycle analysis for molten carbon fuel cells and conventional diesel engines. This research highlighted the environmental benefits of the new power source. Smith et al. [28] suggested that solid oxide fuel cells can be a solution for marine fuel to prevent climate change. However, the study was more or less limited to operational phase. Evrin and Dincer [29] provided thermodynamic analysis and the assessment of an integrated hydrogen fuel cell for ships. In the study, the GHG emission during operation was analyzed. Hansson et al. [30] in the Swedish marine fuel research recommended the hydrogen as the most optimal and the methanol as the second optimal alternative fuels. Nevertheless, LNG and HFO were marked the optimal fuels in the performance and economic criteria.

The literature above provides multi-disciplinary advice for analyzing LCA for marine fuels. On the other hand, it was found that there is still a shortage in case-specific analysis of LNG application to marine vessels when it comes to contributing to developing lifecycle impact of alterative marine fuels. To narrow this gap, this paper was motivated as a preliminary study to evaluate the environmental benefits of using LNG overall by conducting case studies with a newly-constructed LNG-fueled bulk carrier engaged in domestic services of South Korea, one of the world's top five crude oil importers as well as top three LNG consumers in 2018 [31,32]. It is also to introduce a practical approach to evaluate the life cycle emissions from LNG, thereby achieving useful results for the future regulatory framework on the enhanced standardization for maritime emission calculation.

3. Adopted Approach

LCA provides analysis of the environmental prospect and potential environmental impacts throughout life cycle of a product from raw material acquisition through production, use, end-of-life treatment, recycling and final disposal. Economic and social issues are outside the scope of the LCA. The ISO provides standards for the Life Cycle Assessment [33]. Figure 2 displays the approach of the analysis to the LCA framework complying with the ISO standards. It mainly consists of four steps: 'Goal and scope definition', 'Inventory analysis', 'Impact assessment' and 'Interpretation'. The workflow of case studies was assigned in the standardized format.

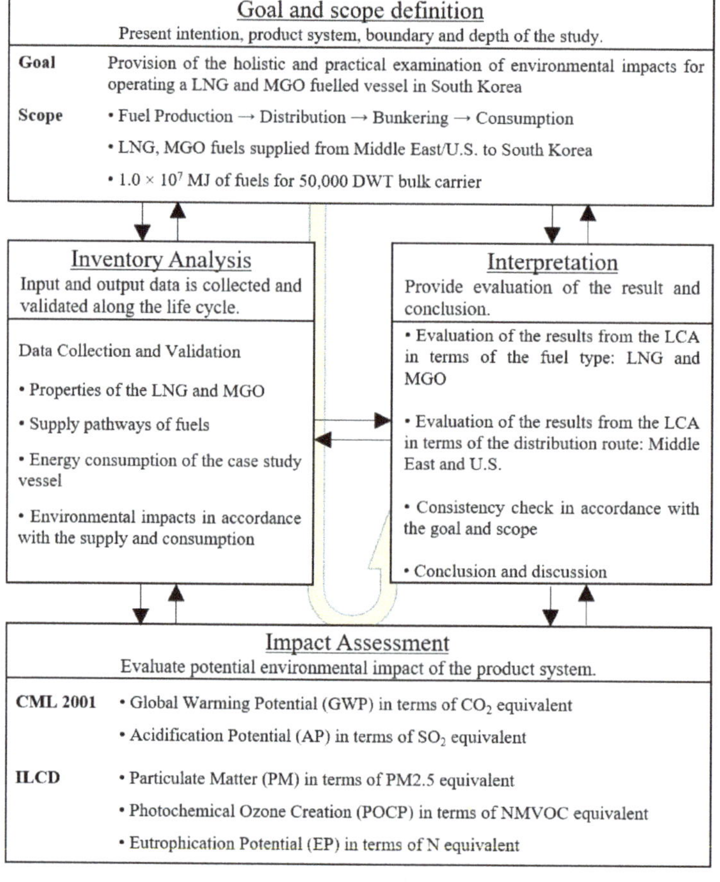

Figure 2. Methodology of Life Cycle Assessment.

3.1. Goal and Scope Definition

The research provides the holistic and practical examination of environmental impacts for operating the LNG- and MGO-fueled vessel in South Korea. The life cycle of the fuels is divided into three phases. Figure 3 illustrates system boundaries of the phases.

1. Well to Tank analysis for emission from the fuel supply
2. Tank to Wake analysis for emission from the fuel combustion
3. Well to Wake analysis for combination of Well to Tank and Tank to Wake analysis

For delivering the extracted natural gas to the final use, several processes and utilities are required. The extracted natural gas is transported using pipelines to purification and liquefaction facilities after production process. The liquefied natural gas is transported by LNG carrier vessels to the LNG terminal and storage facilities of the local industry. The LNG bunker barge or truck delivers the stored LNG to the tank of LNG-fueled ship finally. The stages up to this point are called Well to Tank (WtT) phase. The next stage which means fueled vessel operation is the Tank to Wake (TtW) phase. Otherwise, once the crude oil is extracted, through the production and processing, the crude oil is transported to a refinery in a local oil consuming industry from the oil producing country by pipeline and oil tanker (oil tanker is used only if direct pipeline is not connected between the two region). The refined fuels are delivered to the oil bunkering terminal, and the stored oil in the bunkering terminal is consumed for the ship during voyage at sea.

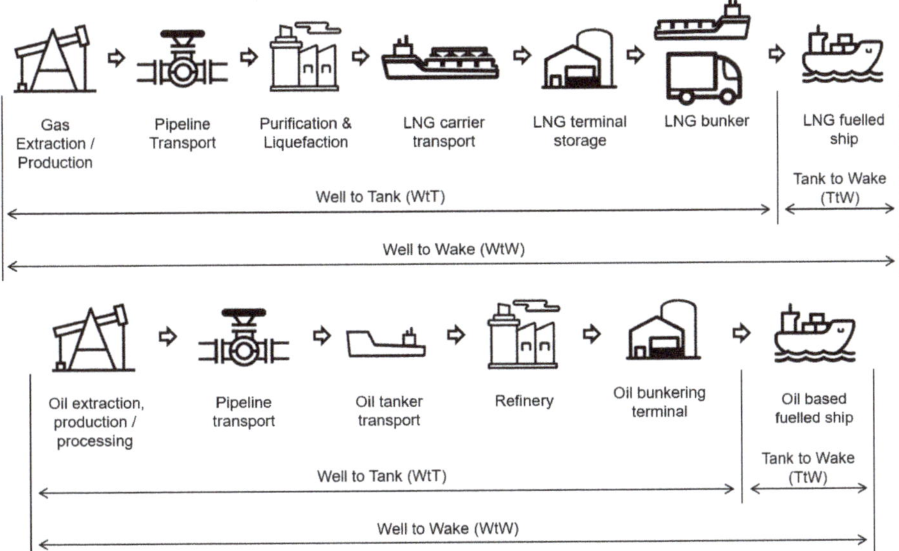

Figure 3. Life cycle of LNG fuel and oil-based marine fuel.

In addition, this research was dealt with under the following conditions.

1. Bunkering operation for the LNG fuel is made by truck-to-ship in the study. Stored LNG fuel in the LNG terminal is transported by means of LNG bunker truck to the port where the vessel is moored. LNG fuel is bunkered directly to the tank of the ship from the tank of the LNG bunker truck [34].
2. Transportations with LNG carriers and LNG bunker trucks include not only laden trips but also ballast trips.

3. Materials and emissions associated with constructing oil and gas facilities (oil extraction, refinery plants, relevant systems, etc.) are not considered in this study. The transport process for energizing the facilities are also excluded.

3.1.1. LNG Supply Chain

Figure 4 represents the LNG imports to South Korea from LNG exporting countries in 2017. Qatar captured the largest amount of LNG fuel as 11.55 Million Tons per Annum (MTPA) which is 31 percentage of the total LNG import in the year [35]. In the same year, the U.S. Energy Information Administration (EIA) reported that South Korea imported approximately 1.99 million tons from U.S. showing 773% of increase compared to the previous year. The long-term contract between South Korea and U.S. validated from 2017 and started import 2.8 MTPA of LNG from Sabine Pass [36].

Considering Korea's energy policy and market share, this research considers two major suppliers of Qatar and the United States. Case 1 refers to LNG import to Korea from Qatar and Case 2 describes LNG import to South Korea from U.S.

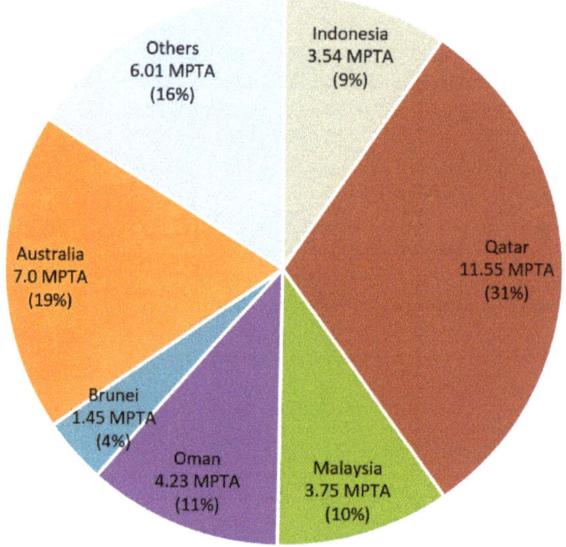

Figure 4. LNG fuel imports to South Korea by exporting countries in 2017.

3.1.2. MGO Supply Chain

South Korea imported approximately 3 million b/d (barrels per day) of crude oil and condensate charting the fifth largest importer in global market. More than 82% of crude oil import was from Middle East [37]. Given this, MGO import from Saudi Arabia to South Korea is assigned to be Case 3. In addition, MGO distribution from U.S. to South Korea is chosen as Case 4 for consistency with the LNG supply point of U.S.

3.1.3. Product System and Functional Unit

WtT analysis represents all steps from fuel extraction to storage in onboard tank. TtW analysis includes actual fuel combustion for ship propulsion. WtT and TtW analyses are integrated to provide a total WtW analysis. Functional unit is defined to be the supply and the consumption of 1.0×10^7 MJ LHV of fuels in consideration of the fuel tanks of a case study vessel which is equipped with 500 m^3 of LNG tank and 400 m^3 of MGO tank. Taking into account the density of LNG, 450 kg/m^3, 225 tons of

LNG is fueled at fully fueled tank [38]. With 48.9 MJ/kg of LHV, LNG fuel provides 1.1×10^7 MJ of energy. 1.0×10^7 MJ LHV of energy is comparable to the available work volume by the case study vessel with the fully fueled LNG tank. In case of MGO, 860 kg/m^3 of density formulates 344 tons of mass at fully tanked condition [39]. With 42.7 MJ/kg of LHV, MGO fuel provides 1.47×10^7 MJ of energy [40]. To provide 1.0×10^7 MJ of energy, 204 tons of LNG is required. In case of MGO, 234 tons of MGO is required for the same energy output. Table 1 summarizes the property of LNG and MGO. The sulphur content of the LNG fuel is assumed to be zero while the MGO fuel contains 0.1% (m/m) of sulphur.

Table 1. Properties and LHV of fuels and required quantity of fuels.

Fuel Type	Tank Size (m^3)	Density (kg/m^3)	Mass of Fuel at Fully Tanked Condition (t)	LHV (MJ/kg)	Total Energy (MJ)	Required Mass of Fuel for 1.0×10^7 MJ (t)
LNG	500	450	225	48.9	1.10×10^7	204
MGO	400	860	344	42.7	1.47×10^7	234

3.2. Inventory Analysis

3.2.1. Well to Tank Inventory Analysis

Transportation pathways of LNG and MGO fuel were determined based on literatures by Korea Gas Corporation (KOGAS), EIA and others [41]. Table 2 summarizes the information on the supply chain while Figure 5 illustrates maritime transportation routes for each case: *Case 1-LNG from Qatar* and *Case 2-LNG from USA* whereas *Case 3-MGO from Saudi Arabia* and *Case 4-MGO from USA*.

Table 2. Design factors for modelling the life cycle of fuels.

Specification	Case 1	Case 2	Case 3	Case 4
Fuel type	LNG	LNG	MGO	MGO
Fuel supplier	Qatar	U.S.	Saudi Arabia	U.S.
Ocean transport distance (period)	23,224 km (30 days)	35,300 km (44 days)	22,902 km (30 days)	36,126 km (44 days)
From fuel terminal to bunkering port	Transported by truck (108 km)		Directly Fueled from oil terminal	
Bunkering operation	Truck-to-Ship		Port-to-Ship	

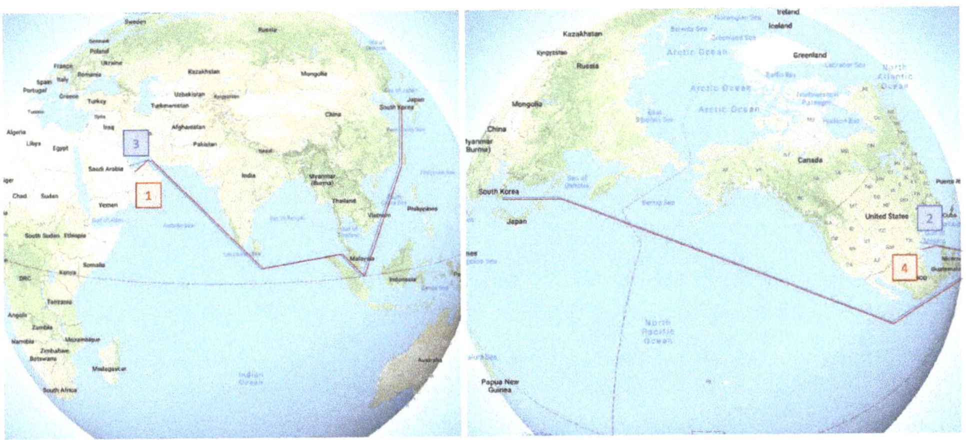

Figure 5. Ocean transport routes of LNG and MGO [42].

Table 3 presents the specification of LNG and crude oil carriers which are considered to transport the marine fuels to bunkering facilities in South Korea. The LNG carrier is considered to use a steam turbine driven by boil-off gas with 30% of efficiency. Boil-off gas is produced 0.15% of LNG cargo per day [13]. Typical types of LNG and crude oil carriers, currently engaged in the service, were selected; the cargo capacity of LNG carrier is 147,237 m^3 whereas crude oil carrier is 57,741 m^3. Table 4 shows the detailed description of the four case studies.

Table 3. Specification of estimated LNG carrier and Crude oil carrier [43].

Ship Type	Engine	Design Speed	Cargo Capacity	Load Factor	Fuel
LNG carrier	27,300 kW	19.5 knot (36.1 km/h)	147,237 m^3	0.55	LNG
Crude oil carrier	12,330 kW	15.2 knot (28.2 km/h)	57,741 m^3	0.55	MGO

Table 4. Descriptions of study cases.

Description	Case 1	Case 2	Case 3	Case 4
Case	*LNG Supply from Qatar to South Korea*	*LNG Supply from U.S. to South Korea*	*MGO Supply from Saudi Arabia to South Korea*	*Oil transportation from U.S to South Korea*
Area of production	North Dome Gas-Condensate field, Qatar	Louisiana gas field, U.S.	Saudi Arabia	Texas oil field, U.S.
Final destination (bunkering port for LNG fuel)	Port of Donghae, South Korea	Port of Donghae, South Korea	Port of Gwangyang, South Korea	Port of Gwangyang, South Korea
Transportation route	North Dome Gas-Condensate field, Qatar → Ras Laffan LNG Terminal, Qatar (Port of Ras Laffan) → Samcheok LNG Terminal, South Korea → Port of Donghae, South Korea	Louisiana gas field, U.S. → Sabine Pass LNG Terminal, U.S. (Port of Cameron LA) → Samcheok LNG Terminal, South Korea → Port of Donghae, South Korea	Ghawar field, Saudi Arabia → Ras Tanura Oil Terminal, Saudi Arabia (Port of Ras Tanura) → Yeosu Oil Terminal (Port of Gwangyang), South Korea	Houston field, U.S. → Houston Fuel Oil Terminal, U.S. (Port of Houston) → Yeosu Oil Terminal (Port of Gwangyang), South Korea
Pipe line length	between North Dome Gas-Condensate field and Ras Laffan LNG terminal: About 85 km [44]	between Louisiana gas field and Sabine Pass LNG Terminal: About 500 km [14]	-	-
Sea route distance	from Ras Laffan LNG Terminal to Samcheok LNG Terminal: About 6270 NM (=11,612 km) for one way voyage [42]	from Sabine Pass LNG Terminal to Samcheok LNG Terminal: 9530 NM (=17,650 km) for one way voyage [42]	from Ras Tanura Oil Terminal to Gwangyang Oil Terminal: 6183 NM (=11,451 km) [42]	from Houston Fuel Oil Terminal to Gwangyang Oil Terminal: 9753 NM (=18,063 km) [42]
Voyage period for vessels	30 days (including return voyage + 2 days for any event)	44 days (including return voyage + 2 days for any event)	30 days (including return voyage + 2 days for any event)	44 days (including return voyage + 2 days for any event)
Truck route distance	from Samcheok LNG Terminal to Port of Donghae: About 54 km for one way trip [45]	from Samcheok LNG Terminal to Port of Donghae: About 54 km for one way trip [45]	-	-
Bunkering operation	Truck-to-Ship	Truck-to-Ship	Port-to-Ship	Port-to-Ship

3.2.2. Tank to Wake Inventory Analysis

The case study vessel is the first LNG-fueled bulker, M/V *Ilshin Green Iris*, which was constructed by Hyundai Mipo Dockyard as a new generation of an environmentally friendly project in July 2016. The vessel obtained a dual class of Lloyd's Register and Korean Register in environmental engineering and complied with the International Gas Fuel (IGF) code [46]. The specification of the vessel is summarized in Table 5.

Table 5. Specification of the case study vessel.

50000 DWT LNG-Fueled Bulk Carrier			
Length overall	190.63 m	**Dead weight scantling**	50,000 MT
Breadth	32.26 m	**Service speed**	14.0 knots (25.9 km/h)
Depth	17.30 m	**Main engine**	2-stroke slow speed diesel
Cargo hold	63,200 m^3	**Daily fuel oil consumption**	21.30 MT/day
LNG fuel tanks	500 m^3	**Daily fuel gas consumption**	17.00 MT/day
MGO tanks	400 m^3	**Cruising range (Oil mode)**	3600 NM
MGO tanks	400 m^3	**Cruising range (Gas mode)**	5300 NM

The vessel is engaged in a regular domestic service in South Korea. Figure 6 illustrates the regular domestic route of the vessel. The sailing route is between the Donghae Port and Gwangyang Port and the distance between the two ports is 271 nautical miles (502 km). Based on the operating profile, four voyages on average are completed each month.

Figure 6. Sea Route of Case Study Vessel [42].

The ship can be operated using three types of fuels: LNG, HFO and MGO. Taking into account IMO sulfur regulation 2020, the LNG and MGO, not HFO, are possible sources of fuel. Emission factors from onboard fuel consumption were determined based on IMO GHG study as shown in Table 6 [47].

Table 6. Emission factors from LNG and MGO consumption. Data from [47].

Emission Factor [kg per 1 kg Consumption of Fuel]								
Fuel Type	**CO_2**	**CO**	**N_2O**	**PM**	**CH_4**	**NO_X**	**NMVOC**	**SO_2**
LNG	2.75	7.83×10^{-3}	1.08×10^{-4}	1.8×10^{-4}	5.0×10^{-2}	1.4×10^{-2}	3.0×10^{-3}	-
MGO	3.21	2.77×10^{-3}	1.6×10^{-4}	9.7×10^{-4}	6.0×10^{-5}	8.7×10^{-2}	3.08×10^{-3}	1.0×10^{-3}

Above table shows that the amount of CO_2 from LNG is 14.3% lower than that of MGO, while LNG produces 833 times more methane (CH_4) than MGO during combustion. Given the significant impact of CH_4 on GWP, it can be perceivable that LNG may have a negative impact on GHG. On the other hand, emissions associated with NO_X and PM from LNG are only about 16.0% and 18.6% of MGO respectively. SO_2, which contributes to AP, is produced from MGO combustion only. The WtW analysis is determined by the combination between the WtT and TtW analyses.

Tables 7 and 8 show the emission data used for the analysis. Those data associated with the LNG production to the LNG carrier transport was adopted from a previous research by [13]. It was found that the production process has contributed to significantly emission of carbon dioxides (CO_2) compared to other pollutants. The same trend was also found in purification and liquefaction processes. CH_4 emission caused by the methane composition in the natural gas from the processes are notable. Methane slip was added to the bunkering operation by 0.0361% [14]. Emissions of the bunker truck are obtained from GaBi database. LNG carrier is considered to use the boil-off gas from the cargo LNG. Since the LNG does not contain sulphur, there is no SO_X emission. Other emissions from the LNG carrier are determined along with the transport distance. Information of LNG terminal storage and bunkering operation was obtained from a research by [20].

Emissions by the MGO supply pathway between well and refinery were referred from a research by [19]. Since crude oil tanker is considered to use MGO fuel, NO_X and SO_X emissions are larger than LNG carrier. Refinery process also produce large amount of SO_X and NO_X emissions rather than LNG purification and liquefaction process. Emissions in MGO terminal storage and bunkering operation are the same as LNG but no methane loss was involved [20].

Table 7. Emissions from LNG supply pathway.

Production & Pipeline Transport [kg per 1 ton of Natural gas]	**CO_2**	**CO**	**N_2O**	**CH_4**	**NO_X**	**S_2O**	**NMVOC**	**PM**
	67.6	0.0917	0.15	0.095	0.428	0.989	0.902	-
Purification & Liquefaction [kg per 1 ton of LNG]	**CO_2**	**CO**	**N_2O**	**CH_4**	**NO_X**	**S_2O**	**NMVOC**	**PM**
	228	0.124	0.00747	1.94	0.187	0.00127	0.0124	0.00124
LNG Carrier Transport, 23,224 km Qatar → Korea [kg per 1 ton of LNG]	**CO_2**	**CO**	**N_2O**	**CH_4**	**NO_X**	**NMVOC**	**PM**	-
	308	0.21	0.00564	0.00564	0.2	0.0139	0.019	-
LNG Carrier Transport, 35,300 km U.S. → Korea [kg per 1 ton of LNG]	**CO_2**	**CO**	**N_2O**	**CH_4**	**NO_X**	**NMVOC**	**PM**	-
	468	0.32	0.00857	0.00857	0.3	0.0211	0.0289	-
LNG Terminal Storage [kg per 1 kg of LNG]	**CO_2**	**CO**	**H_2S**	**NO_X**	**NMVOC**	**SO_X**	-	-
	1.8×10^{-4}	1.3×10^{-6}	1.0×10^{-9}	2.6×10^{-9}	1.1×10^{-5}	4.1×10^{-7}	-	-
Bunker Truck [kg per 108 km transport of 1 kg of LNG]	**Ammonia**	**Benzene**	**CO_2**	**CO**	**PM**	**CH_4**	**NO_2**	**NO**
	1.97×10^{-8}	3.2×10^{-9}	5.72×10^{-3}	3.83×10^{-6}	1.17×10^{-7}	4.59×10^{-9}	8.44×10^{-7}	6.64×10^{-6}
	N_2O	**NMVOC**	**SO_2**	-	-	-	-	-
	3.43×10^{-7}	1.87×10^{-7}	3.60×10^{-8}	-	-	-	-	-
Bunkering Operation [kg per bunkering 1MJ of LNG]	**CO_2**	**CO**	**CH_4**	**H_2S**	**N_2O**	**S_2O**	**NMVOC**	-
	1.8×10^{-4}	1.3×10^{-6}	8.86×10^{-6}	1.0×10^{-9}	2.6×10^{-9}	4.1×10^{-7}	1.1×10^{-5}	-

Table 8. Emissions from MGO supply pathway.

Production & Pipeline Transport [kg per 1 kg of Crude oil]	CO_2	CO	N_2O	NO_X	SO_X	NMVOC	PM
	2.04×10^{-1}	3.33×10^{-5}	2.33×10^{-4}	2.94×10^{-4}	2.51×10^{-6}	3.72×10^{-5}	7.17×10^{-7}
Ocean tanker transportation, 22,902 km Saudi Arabia→ Korea [kg per 1 kg of Crude oil]	CO_2	CO	NO_X	SO_X	NMVOC	PM	
	2.13×10^{-1}	6.23×10^{-4}	5.81×10^{-3}	1.68×10^{-4}	1.73×10^{-4}	1.39×10^{-4}	
Ocean tanker transportation, 36,126 km U.S. → Korea [kg per 1 kg of Crude oil]	CO_2	CO	NO_X	SO_X	NMVOC	PM	
	3.36×10^{-1}	9.83×10^{-4}	9.16×10^{-3}	2.65×10^{-4}	2.73×10^{-4}	2.19×10^{-4}	
Refinery [kg per 1 kg of MGO]	CO_2	CO	N_2O	NO_X	SO_X	NMVOC	PM
	3.00×10^{-1}	3.70×10^{-4}	5.65×10^{-6}	6.99×10^{-4}	1.72×10^{-3}	3.84×10^{-3}	3.26×10^{-5}
MGO Terminal Storage [kg per 1 kg of MGO]	CO_2	CO	H_2S	NO_X	NMVOC	SO_X	
	1.8×10^{-4}	$1.3E \times 10^{-6}$	1.0×10^{-9}	2.6×10^{-9}	1.1×10^{-5}	4.1×10^{-7}	
Bunkering Operation [kg per bunkering 1 MJ of MGO]	CO_2	CO	H_2S	N_2O	S_2O	NMVOC	
	1.8×10^{-4}	$1.3E \times 10^{-6}$	1.0×10^{-9}	2.6×10^{-9}	4.1×10^{-7}	1.1×10^{-5}	

3.2.3. LCA Modelling

While GaBi software was employed for LCA, Figure 7 illustrates the WtT LNG supply model with LNG flow equivalent to the life cycle models of LNG in Figure 3. LNG of 2.04×10^5 kg provides an LHV of 1.0×10^7 MJ, a functional unit for analysis. After the purification and liquefaction process, 2.27×10^5 kg of natural gas is converted into LNG.

The LCA model for MGO supply pathway is presented in Figure 8 which is consistent with the life cycle of oil-based marine fuel as presented in Figure 3. To deliver 1.0×10^7 MJ of MGO in final combustion, 5.52×10^5 kg of crude oil is required. Crude oil is converted to MGO in refineries in South Korea. After the process, 2.34×10^5 kg of MGO is produced, which gives an LHV of 1.0×10^7 MJ as defined by the functional unit.

Figure 9 presents TtW models for both LNG and MGO. This range was limited to onboard fuel consumption. Emissions were calculated based on fuel consumption: 2.04×10^5 kg LNG fuel and 2.34×10^5 kg MGO, giving the same 1.0×10^7 MJ of LHV during combustion.

The LCA model for WtW of LNG and MGO fuels incorporates the Well to Tank and Tank to Wake models as shown in Figure 10.

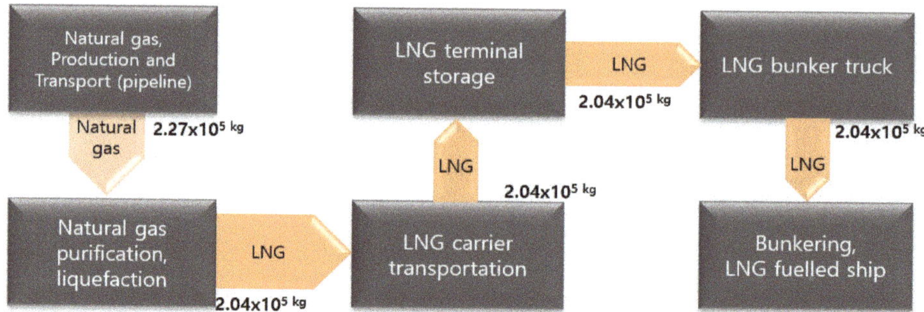

Figure 7. WtT LCA model for LNG fuel.

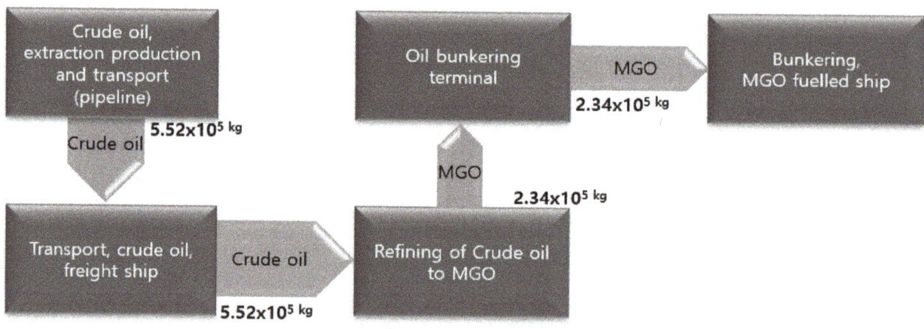

Figure 8. WtT LCA model for MGO fuel.

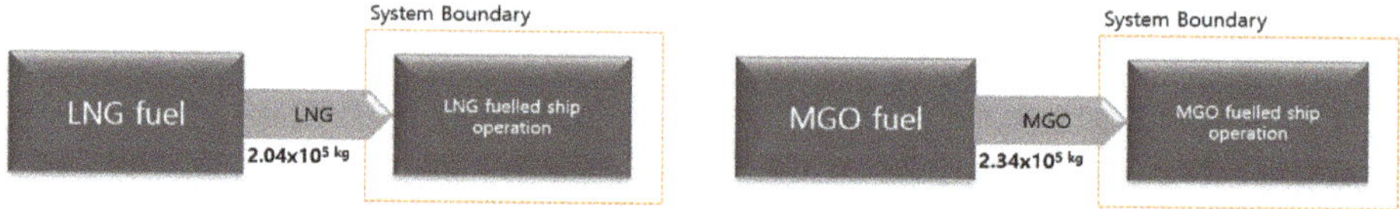

Figure 9. TtW LCA model for LNG and MGO fuel.

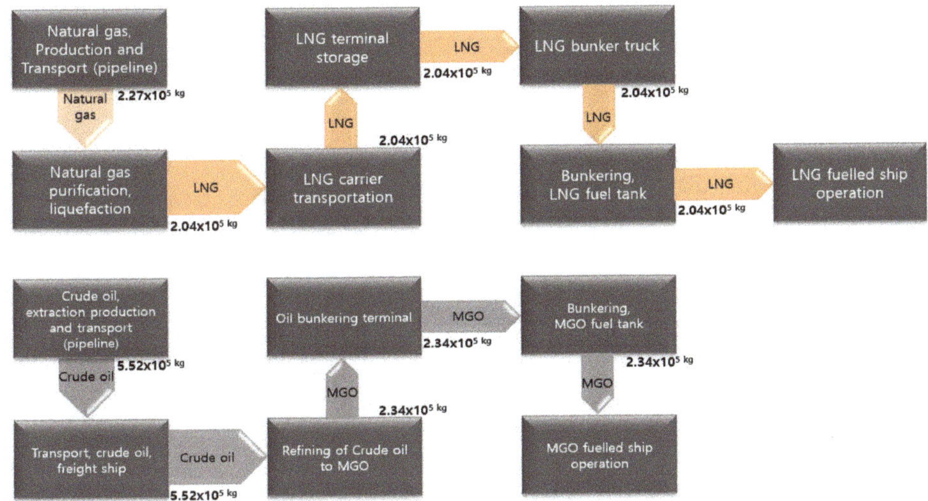

Figure 10. WtW LCA model for LNG fuel and MGO fuel.

3.3. Impact Category

Two kinds of Life Cycle Impact Assessment (LCIA) methodologies are used to assess the impact. GWP in terms of kg CO_2 equivalent and AP in terms of kg SO_2 equivalent are analyzed using CML 2001 method. CML 2001 is midpoint approach method which groups results in midpoint during characterization and normalization [48]. CML 2001 was selected since CML 2001 provides the AP value more visually with kg of SO_2 equivalent particularly.

PM, POCP and EP were analyzed using ILCD method. ILCD is International reference Life Cycle Data System [49]. Since the method provides analysis results for PM, POCP and EP in terms of mass (kg), the ILCD was selected for better-visualized evaluation, PM in terms of kg PM2.5 equivalent, POCP in terms of kg NMVOC equivalent and EP marine in terms of kg N equivalent.

4. Results (Impact Assessment)

4.1. Well to Tank Impact Assessment

The level of GWP from the WtT phase is presented in Figure 11. Since there are no international pipelines in South Korea, the only way to import LNG and MGO fuel is by sea transport. The distances from the resource-export countries determine the sea transport distance and the emissions will vary accordingly.

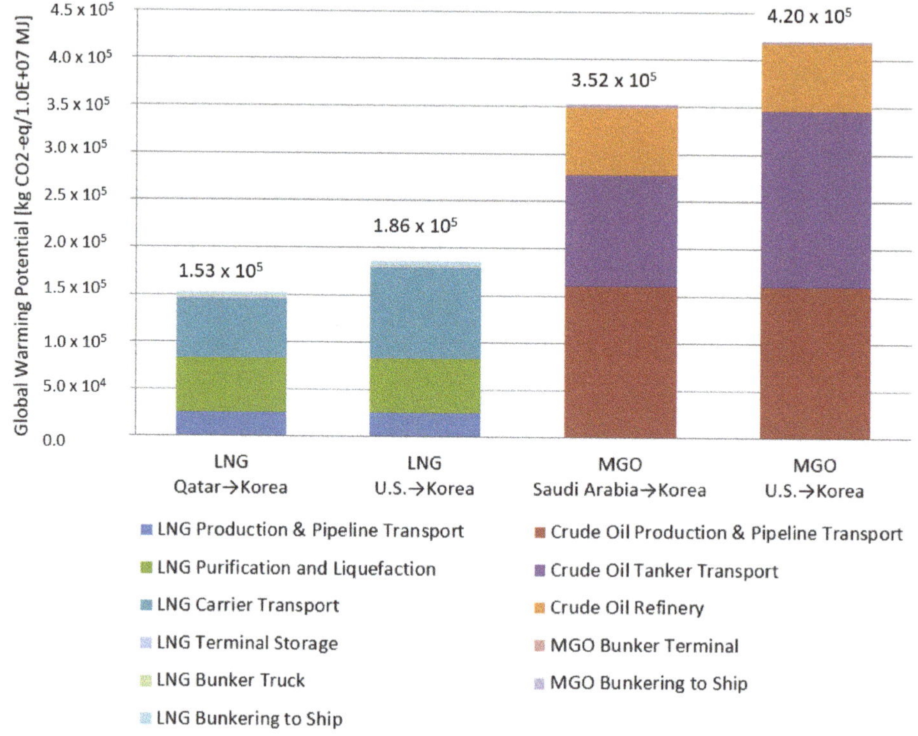

Figure 11. GWP values from supply chain of LNG and MGO.

The simulation results have revealed that emissions from the ocean transport of fuels account for significant portions in this stage. In particular, the LNG carrier contributes to the largest portion of CO_2 equivalent gas in the LNG supply chain. For oil tanker transportation, the route from Saudi Arabia to Korea is not the largest, but the MGO route from the US to Korea produces the largest GWP. LNG supply paths, on the other hand, typically generate less GWP than MGOs.

As explained in the LCA modelling part, crude oil tanker contains whole crude oil than refined MGO. It contributes to the GWP products from the tanker transport; however, it requires to be considered that the crude oil is refined to several energy sources.

Figure 12 illustrates the detail GWP value by LNG supply chain from Qatar to South Korea and MGO supply chain from Saudi Arabia to South Korea for each process.

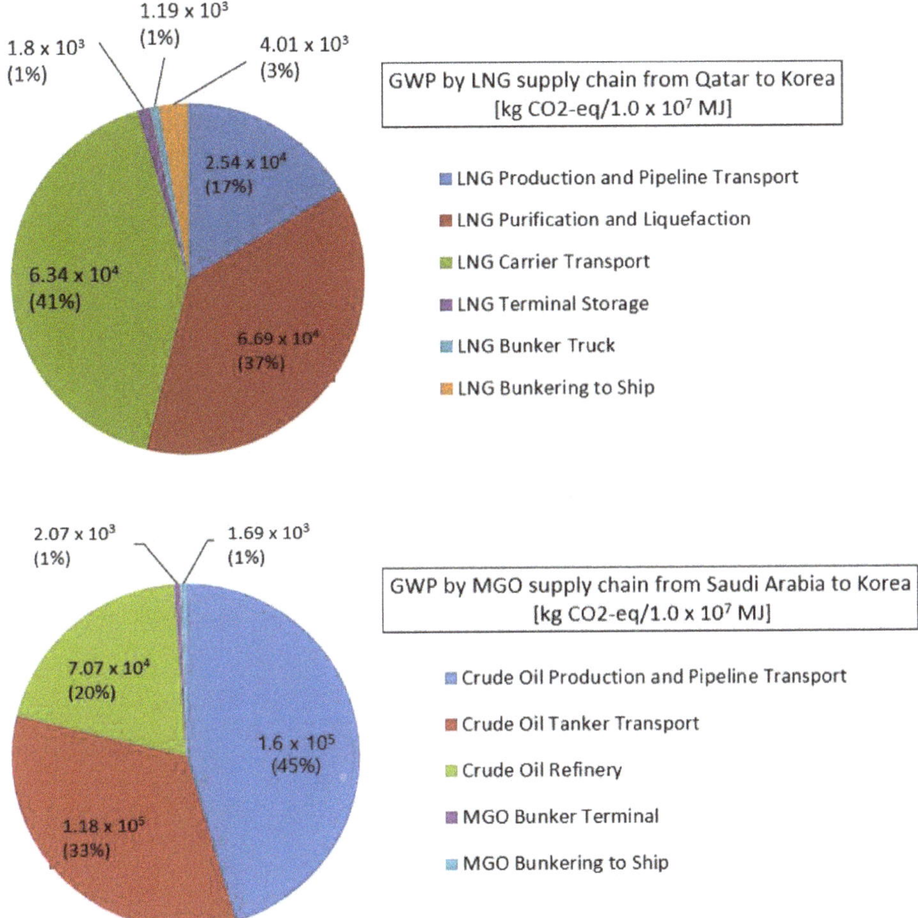

Figure 12. GWP by LNG and MGO supply chain from Middle East to South Korea.

In the LNG supply pathway, next to the ocean transport process, purification and liquefaction process also emits considerable amount of GWP, 5.69×10^4 kg CO_2 equivalent per 1.0×10^7 MJ of LHV. Energy usage and refrigerants losses in these processes contribute substantially to emissions. GHG emissions from utilities for LNG extraction, production and pipelines account for the third part, equivalent to 2.54×10^4 kg CO_2 per 1.0×10^7 MJ LHV.

Diesel trucks emit GHGs from LNG terminals to bunkering ports. The ratio of GWP to total amount is not impressive, but the amount corresponds to more than 1 ton of CO_2.

Methane slip during bunkering operation produces approximately 4 tons of CO_2 equivalent energy per 1.0×10^7 MJ. In case of the MGO fuel, the production and transport process emits a significant amount of GWP, 1.60×10^5 kg CO_2 equivalent per 1.0×10^7 MJ of LHV. Refining operation also produces considerable amount of the greenhouse gas, 7.5×10^4 kg CO_2 equivalent per 1.0×10^7 MJ of LHV. The refining operation also produces a considerable level of GHGs equivalent to 7.5×10^4 kg CO_2 per 1.0×10^7 MJ LHV. AP, PM, POCP and EP results are presented with Figure 12.

The results show that the dominant AP value in the LNG supply chain is attributed to the fuel production stage, whereas the ocean transport of the MGO produces the AP substantially. The result reveals that the dominant AP value in the LNG supply chain is to the fuel production phase. The PM emissions show same aspect with the AP. MGO-Refining process produces the SO_2 equivalent and PM next to the ocean transport. POCP and EP pertinent to the LNG supply chain are relatively lower than the environmental potentials from MGO supply chain. For POCP, ocean transport of the MGO produces the most pollution and the MGO refining process makes a second contribution. Regarding EP, the largest contributor is still ocean transport however production process is second. Referring to the results in Figure 13, ocean transport is shown a key factor to control the local pollutants.

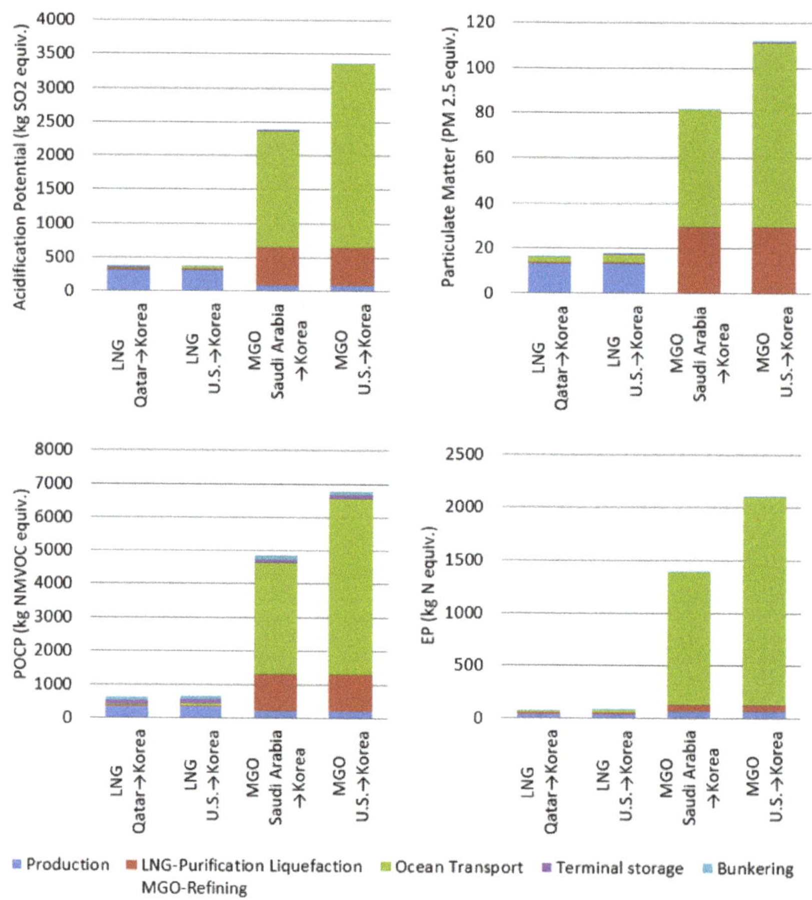

Figure 13. AP, PM, POCP and EP from the case study.

4.2. Tank to Wake Impact Assessment

Figure 14 shows GWP by the ship operation with 1.0×10^7 MJ of the LNG fuel and MGO fuel.

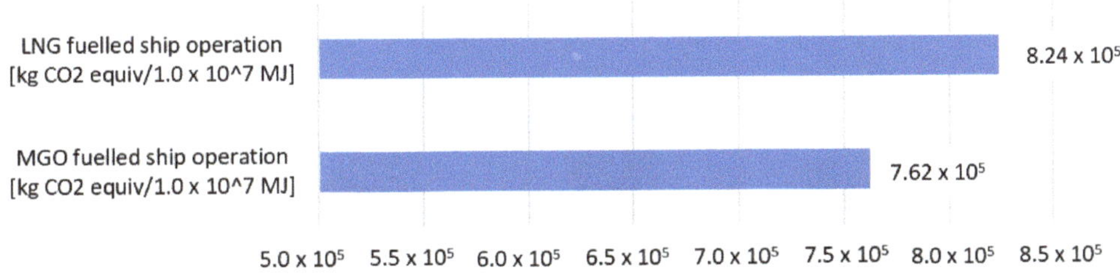

Figure 14. Global Warming Potential from Well to Tank Life Cycle.

GWP by the LNG combustion is approximately 1.08 times larger than MGO fuel's case. In case of LNG fuel combustion, methane slip from LNG fuel consumption raises the value of the CO_2 equivalent considerably. Table 9 exhibits the amount of carbon oxides and methane emissions associated with the on-board fuel combustion process. It is found that the methane emission levels are significantly different in both cases. Although the CO_2 emission is larger at MGO combustion, methane emission expressively contributes total amount of GWP during LNG fuel consumption.

Table 9. Carbon oxides and methane in emissions.

Type	Emission from LNG Consumption [kg]	Emission from MGO Consumption [kg]
Carbon dioxides (CO_2)	5.6×10^5	7.9×10^5
Carbon monoxide (CO)	1.6×10^3	6.89×10^2
Methane (CH_4)	1.0×10^4	1.49×10^1

AP, PM, POCP and EP during fuel consumption are presented in Figure 15. Values of AP, PM, POCP and EP from the MGO fuel consumption are remarkably larger than LNG fuel consumption case. The sulphur content of 0.1% in MGO contributes to the increase in the AP. It is possible to determine that LNG is clean energy based on the analysis results.

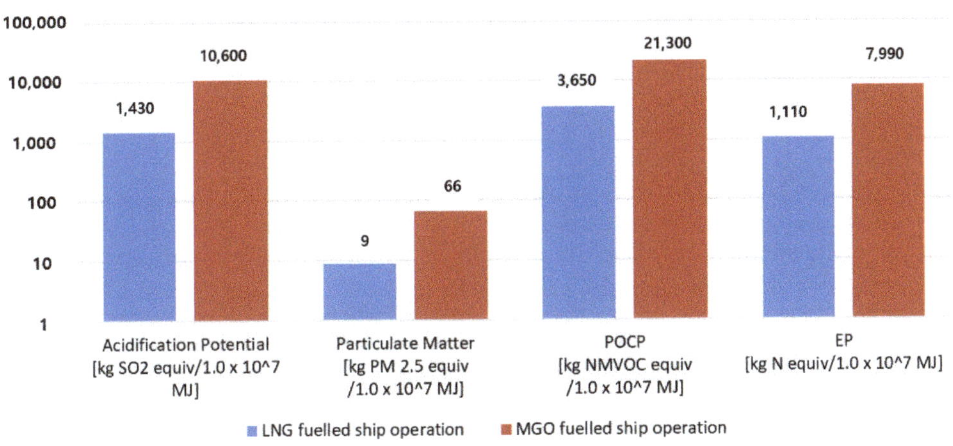

Figure 15. AP, PM, POCP and EP from Fuel Consumption [log scale].

4.3. Well to Wake Impact Assessment

Figure 16 shows the simulation results of the GWP from the WtW phase.

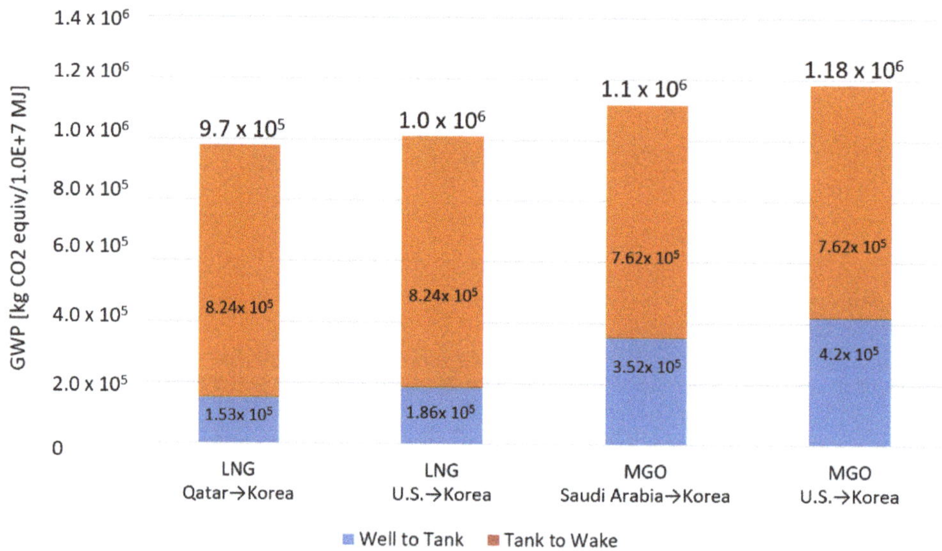

Figure 16. GWP by Well to Wake Analysis.

The overall GWP of MGO in the route of U.S.to South Korea is revealed the largest among the study cases due to the highest levels of the emissions from the supply chain and consumption. GWP during fuel consumption is lower in the MGO case, but the difference between GWPs in the fuel supply chain is much larger. The emission ratio between the supply chain and the onboard consumption is presented in Table 10.

Table 10. Ratio of emissions from the supply chain and consumption.

Fuel Supply Pathway	Share of Emission from Supply Chain [%]	Share of Emission from Consumption [%]	Total [%]
LNG Qatar → Korea	15.7	84.3	100
LNG U.S. → Korea	18.4	81.6	100
MGO Saudi Arabia → Korea	31.6	68.4	100
MGO U.S. → Korea	35.5	64.5	100

Table 10 shows that supply chain emissions are high in the U.S. to South Korea because the sea transport distance of fuel is longer than that in the Middle East to South Korea. In WtT analysis, ocean transport is a significant part of the supply chain.

Figure 17 provides the analysis results for AP, PM, POCP and EP emissions in WtW phase. Except for PM, emissions generated at the WtT stage are more dominant than the TtW stage. In particular, MGOs produce significantly more local pollutants than LNG. The level of GWP is similar between the two fuels, but AP, PM, POCP and EP are about 5–7 times larger in MGO cases than LNG cases.

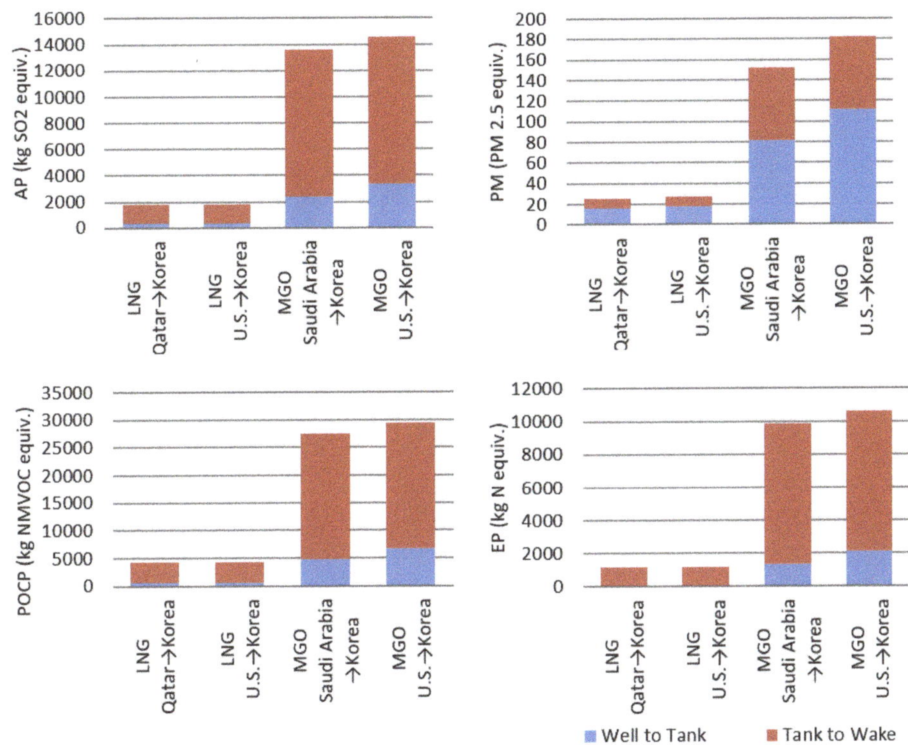

Figure 17. AP, PM, POCP and EP by Well to Wake Analysis.

5. Interpretation and Discussion

Although both LNG and MGO fuels are able to meet IMO 2020 sulphur regulations, the environmental impacts of those fuels have been found to significantly vary depending on fuel type and supply chain. Given this, the modelling of each life stage of the fuels has been shown to help identifying the potential processes where we can optimize and reduce emission levels. In this research, the environmental impact of the processes pertinent to the fuel supply chain was proven significantly high. Hence, several ways to reduce emission levels by optimizing transportations can be proposed as follows:

1. Application of international pipelines or other cleaner transport solutions;
2. Ensuring closest LNG exporter;
3. Considering FLNG (Floating Liquefied Natural Gas);
4. Using renewable energy for domestic transports;
5. Domestic pipelines for fuel distribution.

On the other hand, the WtW analysis shows that the emission levels from onboard fuel consumption are dominant in the overall process. Although CO_2 emissions from LNG combustion are lower than the MGO case, simulation results from the TtW impact assessment indicate that methane slip from LNG fuel engine operation is an important factor for increasing GWP emissions. It reveals that the life cycle GWP level from the LNG is 9.77×10^5 kg CO_2 equivalent per 1.0×10^7 MJ fuel consumption, whereas that from the MGO is 1.11×10^6 kg CO_2 equivalent per 1.0×10^7 MJ fuel consumption. These results provide insight into why the marine industry should strive to minimize methane slip from engine combustion.

In short, this research demonstrated the excellence of using LNG as a marine fuel relative to the MGO. However, it cannot neglect emphasizing on the need to resolve the methane slip issues; if successful solutions arrive in the marine industry, the use of LNG as a marine fuel can further contribute to reduction of the GHG emissions.

Results of the impact assessment for local pollutants even more clearly support the superiority of LNG over MGO since the MGO fuel cases are revealed to contribute 5–7 times higher emissions than the LNG cases. In this respect, the LNG is confidently recommended over the MGO as long as we are solely concerned with the environmental perspective.

On the other hand, the demands on addressing the shortcomings of current maritime environmental indicators, EEDI and EEOI, are increasing. Given this, the life cycle assessment can be an effective approach to estimate holistic emission levels attributed by marine fuels from the extraction to the final use. Taking into account that the marine industry is grappling with lowering emission levels more comprehensively, this research is highly believed to be an important primary study that will help to provide practical solutions to assess the all-inclusive environmental impacts of marine fuels. Such solutions will help us to obtain a clear knowledge on the holistic environmental impacts from shipping. To achieve this goal, the future research is recommended to expand the case studies to various alternative fuel types and supply chains to determine the most optimal future marine fuel as well as to obtain general observations on the level of differences and how to improve it.

6. Conclusions

The key research findings can be summarized as below.

Using LNG as a marine fuel was proven as an effective fuel in reducing the marine pollutant, compared to conventional marine petroleum fuels including MGO. In particular, Case 1 (LNG import from Qatar) was revealed the best option based on that all environmental impact categories were marked the lowest, compared to other cases. The emission quantities for Case 1 were presented as below:

(1) GWP: 977 tonnages of CO_2 equivalent per 1.0×10^7 MJ of fuel consumption;

   1. AP: 1.76 tonnages of SO_2 equivalent per 1.0×10^7 MJ of fuel consumption;
   2. EP: 1.18 tonnages of N equivalent per 1.0×10^7 MJ of fuel consumption;
   3. POCP: 4.28 tonnages of NMVOC equivalent per 1.0×10^7 MJ of fuel consumption;
   4. PM: 26 kg of $PM_{2.5}$ equivalent per 1.0×10^7 MJ of fuel consumption.

   Moreover, considering factor-based findings, Case 1, LNG from Middle East to South Korea, was determined to produce 0.88 times lesser GWP compared to Case 3, MGO from Middle East to South Korea while Case 2, LNG from U.S. to South Korea, emitted 0.86 times lesser GWP than Case 4, MGO from U.S. to South Korea.

(2) Regional distances between the energy exporter and the importer and the supply chains were found to be important parameters to determine the environmental impact of marine fuels, which suggests the importance of optimal production, transport as well as usage.

(3) The methane slip pertinent to LNG combustion was identified an issue to be resolved in order to adopt the LNG as a successful post-2020 marine fuel; the amount of GWP contributed by the methane slip was not negligibly small.

(4) LCA was proved effective for marine industry including oil and gas as a comprehensive and robust tool to evaluate the holistic environmental impact on marine pollutions. This proposed approach is believed to contribute to addressing the shortcomings of current maritime emission calculators. Therefore, the analysis results and the proposed approach are to provide the stakeholders an insight into making proper decision-making and future regulatory framework.

Author Contributions: Conceptualization, B.J. and S.H.; methodology, B.J., S.H., and M.K.; software, S.H.; formal analysis, investigation, S.H.; resources, B.J. and S.H.; writing, B.J., S.H., and M.K.; review and editing, M.K. and P.Z.; visualization, S.H.; supervision, M.K., K.J., P.Z.; funding acquisition, K.J.; project administration, B.J. and K.J.

Funding: This research is supported by Pusan National University (PNU) Korea-UK Global Program in Offshore Engineering (N0001288) funded by the Korean Ministry of Trade, Industry and Energy.

Acknowledgments: The authors would like to express their gratitude to ILSHIN shipping CO., Ltd. and Korean Register of Shipping for their invaluable support, comments and suggestions. They have contributed considerably to this study.

Conflicts of Interest: The authors declare no conflict of interest.

Abbreviations

AP	Acidification Potential
DWT	Dead Weight Tonnage
ECA	Emission Control Area
EP	Eutrophication Potential
FLNG	Floating Liquefied Natural Gas
GHG	Green House Gas
GWP	Global Warming Potential
HFO	Heavy Fuel Oil
HSFO	High Sulphur Fuel Oil
IMO	International Maritime Organization
ISO	International Organization for Standardization
LCA	Life Cycle Assessment
LCI	Life Cycle Inventory
LHV	Lower Heating Value
LNG	Liquefied Natural Gas
LSFO	Low Sulphur Fuel Oil
MGO	Marine Gas Oil
MTPA	Million Tonnes per Annum
NGL	Natural Gas Liquid
NMVOC	Non-Methane Volatile Organic Compounds
NO_X	Nitrogen Oxides
PM	Particulate Matter
POCP	Photochemical Ozone Creation Potential
SO_X	Sulphur Oxides
TtW	Tank to Wake
ULSFO	Ultra-Low Sulphur Fuel Oil
WtT	Well to Tank
WtW	Well to Wake

References

1. UN. *Review of Marine Transport 2018*; UNCTAD: Geneva, Switzerland, 2018.
2. Corkhill, M. LNG World Shipping—Healthy Freight Rates and 2018 LNG Shipping Predictions. 2018. Available online: https://www.lngworldshipping.com/news/view,healthy-freight-rates-and-2018-lng-shipping-predictions_50323.htm (accessed on 31 August 2018).
3. Yoo, B.Y. Economic assessment of liquefied natural gas (LNG) as a marine fuel for CO2 carriers compared to marine gas oil (MGO). *Energy* **2017**, *121*, 772–780. [CrossRef]
4. Piellisch, R. 'Container Ship LNG? Plenty of Time', HHP Insight. 2013. Available online: http://hhpinsight.com/marine/2013/01/container-ship-lng-plenty-of-time/ (accessed on 18 January 2016).
5. Thomson, H.; Corbett, J.J.; Winebrake, J.J. Natural gas as a marine fuel. *Energy Policy* **2015**, *87*, 153–167. [CrossRef]
6. El-Gohary, M.M. The future of natural gas as a fuel in marine gas turbine for LNG carriers. *Proc. Inst. Mech. Eng. Part M J. Eng. Marit. Environ.* **2012**, *226*, 371–377. [CrossRef]

7. IMO. *Guidelines on the Method of Calculation of the Attained Energy Efficiency Design Index (EEDI) for New Ships*; International Maritime Organization: London, UK, 2018.
8. IMO. *ISWG-GHG 5/4/5: Consideration of Concrete Proposals on Candidate Short-Term Measures*; Consideration on Development of the Lifecycle GHG/Carbon Intensity Guidelines for all Types of Fuels; International Maritime Organization: London, UK, 2019.
9. Guinée, J.B. Handbook on life cycle assessment operational guide to the ISO standards. *Int. J. Life Cycle Assess.* **2002**, *7*, 311–313. [CrossRef]
10. Finnveden, G.; Hauschild, M.Z.; Ekvall, T.; Guinée, J.; Heijungs, R.; Hellweg, S.; Koehler, A.; Pennington, D.; Suh, S. Recent developments in Life Cycle Assessment. *J. Environ. Manag.* **2009**, *91*, 1–21. [CrossRef] [PubMed]
11. Levasseur, A.; Lesage, P.; Margni, M.; Deschenes, L.; Samson, R. Considering Time in LCA: Dynamic LCA and Its Application to Global Warming Impact Assessments. *Environ. Sci. Technol.* **2010**, *44*, 3169–3174. [CrossRef]
12. Woods, J.S.; Veltman, K.; Huijbregts, M.A.; Verones, F.; Hertwich, E.G. Towards a meaningful assessment of marine ecological impacts in life cycle assessment (LCA). *Environ. Int.* **2016**, *89*, 48–61. [CrossRef]
13. Bengtsson, S.; Andersson, K.; Fridell, E. *Life Cycle Assessment of Marine Fuels: A Comparative Study of Four Fossil Fuels for Marine Propulsion*; Chalmers University of Technology: Gothenburg, Sweden, 2011.
14. Thinkstep. *Life Cycle GHG Emission Study on the Use of LNG as Marine Fuel*; 2019. Available online: www.thinkstep.com/content/life-cycle-ghg-emission-study-use-lng-marine-fuel-1 (accessed on 23 June 2019).
15. El-Houjeiri, H.; Monfort, J.-C.; Bouchard, J.; Przesmitzki, S. Life Cycle Assessment of Greenhouse Gas Emissions from Marine Fuels: A Case Study of Saudi Crude Oil versus Natural Gas in Different Global Regions. *J. Ind. Ecol.* **2019**, *23*, 374–388. [CrossRef]
16. Sharafian, A.; Blomerus, P.; Mérida, W. Natural gas as a ship fuel: Assessment of greenhouse gas and air pollutant reduction potential. *Energy Policy* **2019**, *131*, 332–346. [CrossRef]
17. Liu, R.E. Life Cycle Greenhouse Gas Emissions of Western Canadian Natural Gas and a Proposed Method for Upstream Life Cycle Emissions Tracking; 2019. Available online: http://hdl.handle.net/1880/110466 (accessed on 29 July 2019).
18. Brynolf, S.; Fridell, E.; Andersson, K. Environmental assessment of marine fuels: Liquefied natural gas, liquefied biogas, methanol and bio-methanol. *J. Clean. Prod.* **2014**, *74*, 86–95. [CrossRef]
19. Tan, R.B.; Wijaya, D.; Khoo, H.H. LCI (Life cycle inventory) analysis of fuels and electricity generation in Singapore. *Energy* **2010**, *35*, 4910–4916. [CrossRef]
20. Tagliaferri, C.; Clift, R.; Lettieri, P.; Chapman, C. Liquefied natural gas for the UK: A life cycle assessment. *Int. J. Life Cycle Assess.* **2017**, *49*, 1944–1956. [CrossRef]
21. Jeong, B. Comparative Analysis of SO_x Emission-Compliant Options for Marine Vessels from Environmental Perspective. *J. Korean Soc. Power Syst. Eng.* **2018**, *22*, 72–78. [CrossRef]
22. Hua, J.; Cheng, C.-W.; Hwang, D.-S. Total life cycle emissions of post-Panamax containerships powered by conventional fuel or natural gas. *J. Air Waste Manag. Assoc.* **2019**, *69*, 131–144. [CrossRef]
23. Jeong, B.; Jang, H.; Zhou, P.; Lee, J.U. Investigation on marine LNG propulsion systems for LNG carriers through an enhanced hybrid decision making model. *J. Clean. Prod.* **2019**, *230*, 98–115. [CrossRef]
24. Rocco, M.V.; Langè, S.; Pigoli, L.; Colombo, E.; Pellegrini, L.A. Assessing the energy intensity of alternative chemical and cryogenic natural gas purification processes in LNG production. *J. Clean. Prod.* **2019**, *208*, 827–840. [CrossRef]
25. Miksch, T.P. USA-Asia LNG Shipping Route Optimization. Master Thesis, University Putra Malaysia (UPM), Selangor Darul Ehsan, Malaysia, February 2019.
26. Dong, D.T.; Cai, W. Life-cycle assessment of ships: The effects of fuel consumption reduction and light displacement tonnage. *Proc. Inst. Mech. Eng. Part M J. Eng. Marit. Environ.* **2019**. [CrossRef]
27. Alkaner, S.; Zhou, P. A comparative study on life cycle analysis of molten carbon fuel cells and diesel engines for marine application. *J. Power Sources* **2006**, *158*, 188–199. [CrossRef]
28. Smith, L.; Ibn-Mohammed, T.; Yang, F.; Reaney, I.M.; Sinclair, D.C.; Koh, S.L. Comparative environmental profile assessments of commercial and novel material structures for solid oxide fuel cells. *Appl. Energy* **2019**, *235*, 1300–1313. [CrossRef]
29. Evrin, R.A.; Dincer, I. Thermodynamic analysis and assessment of an integrated hydrogen fuel cell system for ships. *Int. J. Hydrogen Energy* **2019**, *44*, 6919–6928. [CrossRef]

30. Hansson, J.; Månsson, S.; Brynolf, S.; Grahn, M. Alternative marine fuels: Prospects based on multi-criteria decision analysis involving Swedish stakeholders. *Biomass Bioenergy* **2019**, *126*, 159–173. [CrossRef]
31. Workman, D. Croude Oil Imports by Country. 2019. Available online: http://www.worldstopexports.com/crude-oil-imports-by-country/ (accessed on 25 June 2019).
32. EIA. China Becomes World's Second Largest LNG Importer, Behind Japan. 2019. Available online: https://www.eia.gov/todayinenergy/detail.php?id=35072 (accessed on 25 June 2019).
33. Finkbeiner, M.; Inaba, A.; Tan, R.; Christiansen, K.; Klüppel, H.-J. The New International Standards for Life Cycle Assessment: ISO 14040 and ISO 14044. *Int. J. Life Cycle Assess.* **2006**, *11*, 80–85. [CrossRef]
34. Lee, C. Status and Prospects of LNG Bunkering by KOGAS. In Proceedings of the International LNG Fueled Ship & Bunkering Conference, Busan, Koran, 15 November 2018.
35. KESIS. Annual LNG Import Report. 2018. Available online: http://www.kesis.net/sub/subChart.jsp?report_id=7020200&reportType=0 (accessed on 24 July 2019).
36. Paik, K.-W. South Korea's Energy Policy Change and the Implications for Its LNG Imports. Available online: https://doi.org/10.26889/9781784671136. (accessed on 15 June 2019).
37. U.S.EIA. *Country Analysis Brief: South Korea*; U.S. Energy Information Administer: Washington, DC, USA, 2018.
38. Thinkstep. GaBi Database. Available online: http://www.gabi-software.com/international/databases/gabi-databases. (accessed on 15 June 2019).
39. GEOS, G. A Guide to Fuel Properties. 2014. Available online: http://www.geosgroup.com/news/article/a-guide-to-fuel-properties (accessed on 24 July 2019).
40. Vermeire, M.B. *Everything You Need to Know about Marine Fuels*; Chevron Global Marine Products: Singapore, 2012.
41. KOGAS. *Technical Reports for New LNG Carrier Construction: LNG*; Korea Gas Corporation: Daegu, Korea, 2013. (In Korean)
42. SEAROUTES. Available online: http://www.searoutes.com (accessed on 13 June 2019).
43. Knaggs, T. *Significant Ships of 2008*; RINA: London, UK, 2008.
44. Naji, A.-A. Natural Gas Reserves, Development and Production in Qatar. 1998. Available online: https://inis.iaea.org/collection/NCLCollectionStore/_Public/29/029/29029263.pdf (accessed on 23 June 2019).
45. NAVER. Available online: https://map.naver.com/ (accessed on 23 June 2019).
46. World Maritime News. *World's First LNG-Fueled*; World Maritime News: Schiedam, The Netherlands, 2019.
47. I.M.O. *Greenhouse Gas Study, Third, 2014*; Executive Summary: London, UK, 2014.
48. Dreyer, L.C.; Niemann, A.L.; Hauschild, M.Z. Comparison of Three Different LCIA Methods: EDIP97, CML2001 and Eco-indicator 99. *Int. J. Life Cycle Assess.* **2003**, *8*, 191–200. [CrossRef]
49. Commission, E.U.-E. *International Reference Life Cycle Data System (ILCD) Handbook—General Guide for Life Cycle Assessment—Detailed Guidance*; Institute for Environment and Sustainability: Ispra, Italy, 2010.

MDPI
St. Alban-Anlage 66
4052 Basel
Switzerland
Tel. +41 61 683 77 34
Fax +41 61 302 89 18
www.mdpi.com

Journal of Marine Science and Engineering Editorial Office
E-mail: jmse@mdpi.com
www.mdpi.com/journal/jmse